Professional Practice Cases in Water Supply and Drainage Science and Engineering

给排水科学与工程专业实习指导

杜海霞　吴慧芳　主编
Haixia Du　Huifang Wu

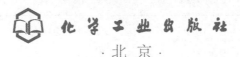

北　京

Synopsis

The textbook is designed for the main problems involved in the practice of students in Water Supply and Prainage Science and Engineering, so that students can better practice with questions. The content contains utilization and protection of water resources, water intake, distribution and delivery facilities, traditional water treatment process, combined and separate sewer, wastewater pretreatment, primary wastewater treatment system and secondary wastewater treatment system, sludge treatment and disposal, and sewage advanced treatment system.

The textbook focuses on the principle of the overall knowledge of water supply and drainage, the key points, problems and precautions in the actual investigation and practice of students. We emphasizes the importance of the design of teaching materials for the practical training of water supply and drainage. And the textbook combines the latest improvement on our understanding of basic phenomena as well as research and application of new technologies.

图书在版编目（CIP）数据

给排水科学与工程专业实习指导＝Professional Practice Cases in Water Supply and Drainage Science and Engineering：英文/杜海霞，吴慧芳主编. —北京：化学工业出版社，2020.5

ISBN 978-7-122-35599-7

Ⅰ.①给… Ⅱ.①杜… ②吴… Ⅲ.①给排水系统-高等学校-教材-英文 Ⅳ.①TU991

中国版本图书馆 CIP 数据核字（2020）第 079206 号

责任编辑：满悦芝　　　　　　　　　　文字编辑：吴开亮
责任校对：张雨彤　　　　　　　　　　装帧设计：张　辉

出版发行：化学工业出版社（北京市东城区青年湖南街 13 号　邮政编码 100011）
印　　装：涿州市京南印刷厂
710mm×1000mm　1/16　印张 10¾　字数 176 千字　2020 年 9 月北京第 1 版第 1 次印刷

购书咨询：010-64518888　　　　　　　售后服务：010-64518899
网　　址：http://www.cip.com.cn
凡购买本书，如有缺损质量问题，本社销售中心负责调换。

定　　价：49.90 元　　　　　　　　　　　　　　　　　　　版权所有　违者必究

Preface

Based on the current scarcity of practical English teaching materials for water supply and drainage science and engineering majors, this textbook is written for the main problems and precautions involved in the practice and training of students, so that students can better practice with questions.

The textbook combines the principles of instrumentality, innovation, practice, knowledge and the combination of inside and outside the class. With a gradual structure, the textbook follows the cognitive rules of students, and organically combines the mutual penetration mode of in-class learning and extracurricular practice, internalized the knowledge learned in the class into application ability and good habits of students through extracurricular practice.

At the same time, in order to improve the international competitiveness of students in cultural literacy, professional skills and other aspects, the textbook is written in English to provide reference and guidance for both in-class teaching and extracurricular practice in English on basic principles, key points to be paid attention to in social investigation and field practice, and problems to be considered in field operation practice and precautions. The content contains utilization and protection of water resources, water intake, distribution and delivery facilities, traditional water treatment process, combined and separate sewer, wastewater pretreatment, primary wastewater treatment system and secondary wastewater treatment system, sludge treatment and disposal, and sewage advanced treatment system.

We would like to express our great appreciation to Prof. Fusheng Li, Prof. Shujuan Zhang and Prof. Yongjun Sun for all kindness assistants and valuable informations, suggestions and comments during the process of writing the textbook.

Haixia Du, Huifang Wu
2020. 6

Contents

Chapter 1   Introduction   1

    1.1 Water supply   1
    1.2 Sewerage   1

Chapter 2   Water use and water quality   3

    2.1 Relation of water quantity and population   4
    2.2 Water use for different purposes   7
    2.3 Water impurities   7
    2.4 Waterborne diseases   8
    2.5 Inorganic contaminants   9
    2.6 Organic contaminants   11
    2.7 Drinking water quality standards of China   12
    2.8 Drinking water quality standards of Japan   18
    2.9 US EPA (Environmental Protection Agency) water quality standards   22

Chapter 3   Water sources and water intake facilities   30

    3.1 Rainwater   31
    3.2 Surface water   34
    3.3 Groundwater   37
    3.4 Reservoir storage   40
    3.5 Water intake facilities   41

Chapter 4   Conventional drinking water treatment processes   49

    4.1 Conventional rapid sand filtration system   49
    4.2 Coagulation and flocculation   50

4.3 Sedimentation 52

4.4 Rapid sand filtration 55

4.5 Disinfection 57

Chapter 5 Water distribution 60

5.1 Distribution reservoirs 61

5.2 Distribution pipe system 64

5.3 Pumping stations 65

Chapter 6 Miscellaneous water treatment techniques 69

6.1 Advanced oxidation (ozonation, UV radiation and hydroxyl oxidation) 69

6.2 Activated carbon adsorption 76

6.3 Ion exchange 80

6.4 Membrane filtration 83

6.5 Reverse osmosis 86

6.6 Biological filtration 88

Chapter 7 The sewers 93

7.1 Combined sewers 93

7.2 Separate sewers 96

Chapter 8 Characteristics of wastewater 100

8.1 Organic substances 100

8.2 Inorganic substances 101

8.3 Pathogenic microorganisms 102

Chapter 9 Preliminary wastewater treatment systems 104

9.1 Screening 104

9.2 Flow equalization tank 105

9.3　Grit clarifier　106

Chapter 10　Primary wastewater treatment systems　　109

10.1　Plain sedimentation　109

10.2　Aerated sedimentation　111

Chapter 11　Secondary wastewater treatment systems　　113

11.1　Conventional activated sludge process　114

11.2　Oxidation ditch process　117

11.3　Trickling filter process　119

11.4　Anaerobic waste treatment processes　123

Chapter 12　Sludge treatment and disposal　　129

12.1　Characteristics of sludge　130

12.2　Sludge conditioning, thickening and dewatering　134

12.3　Digestion　139

12.4　Composting　143

12.5　Incineration　146

12.6　Microbial fuel cell　147

Chapter 13　Advanced wastewater treatment　　151

13.1　Nitrogen removal　152

13.2　Phosphorus removal　158

References　　162

Chapter 1 Introduction

Water is a precious resource and vital for life. Without it we would die within days. Access to a safe and affordable supply of drinking water is universally recognized as a basic human need for the present generation and a pre-condition for the development and care of the next. Water is also a fundamental economic resource on which people's livelihoods depend. In addition to domestic water use, households use water for productive activities such as farming and livestock rearing in rural areas, or horticulture and home-based micro-enterprises in urban settlements.

1.1 Water supply

Water supplies for agriculture, industry, power generation, ecosystem protection, navigation, etc., involve different considerations. There are evident and important links between domestic water supply/sanitation and the management of water resources as a whole. Though water for domestic use accounts for only about five percent of water consumption, it is a proportion that must be safeguarded in both quality and quantity as a basic human need. At the same time, poor sanitation practices are the major cause of surface and groundwater pollution.

1.2 Sewerage

Urban drainage includes the removal of all unwanted water from urban areas. It includes wastewater-including sewerage, grey water and stormwater. Grey water is domestic wastewater predominately from baths, basins and washing machines. The unwanted water may be used for other purposes with or without treatment.

Sewerage is the infrastructure that conveys sewage or surface runoff (stormwater, meltwater and rainwater). It encompasses components such as receiving drains, manholes, pumping stations, storm overflows, and screening chambers of the combined sewer or sanitary sewer. Sewerage ends at the entry to a sewage treatment plant or at the point of discharge into the environment. It is the system of pipes, chambers, manholes, etc., that conveys the sewage or storm water. According to this definition, sewerage and sewage are two different terms. However, at least in American English colloquial usage, the both terms are sometimes used interchangeably.

Chapter 2 Water use and water quality

It is estimated that 70% of worldwide water is used for irrigation, with 15%-35% of irrigation withdrawals being unsustainable. It takes around 2000-3000L of water to produce enough food to satisfy one person's daily dietary need. This is a considerable amount, when compared to that required for drinking, which is between 2L and 5L. 22% of worldwide water is used in industry. Major industrial users include hydroelectric dams, thermoelectric power plants (using water for cooling), ore and oil refineries (using water in chemical processes), and manufacturing plants (using water as a solvent). Water withdrawal can be very high for certain industries, but consumption is generally much lower than that of agriculture. 8% of worldwide water use is for domestic purposes. These include drinking water, bathing water, cooking water, toilet flushing water, cleaning water, laundry water and gardening water. Basic domestic water requirements have been estimated at around 50L per person per day, excluding water for gardens. Drinking water is the water that is of sufficiently high quality so that it can be consumed or used without risk of immediate or long term harm. Such water is commonly called potable water. In most developed countries, the water supplied to domestic, commerce and industry is all of drinking water standard even though only a very small proportion is actually consumed or used in food preparation.

There are numerous measures of water use, including total water use, drinking water consumption, non-consumptive use, withdrawn water use (from surface and groundwater sources), in-stream use, water footprint, etc. Each of these measures of water use is appropriate for some purposes and inappropriate for others. Water "footprints" have become popular measures of use, e. g. in relation to personal consumption. The term "water footprint" is often used to refer to the amount of water used by an individual, community, business, or nation, or the amount of water use associated with (although not necessarily assignable to) a product.

Water quality refers to the chemical, physical, biological and radiological characteristics of water. It is a measure of the condition of water relative to the requirements of one or more biotic species and or to any human need or purpose. It is most frequently used by reference to a set of standards against which compliance can be assessed. The most common standards used to assess water quality relate to health of ecosystems, safety of human contact, and drinking water.

2.1 Relation of water quantity and population

2.1.1 Fundamentals

In 2000, the world population was 6.2 billion. The UN estimates that by 2050 there will be an additional 3.5 billion people with most of the growth in developing countries that already suffer water stress. Thus, water demand will increase unless there are corresponding increases in water conservation and recycling of this vital resource. when the population grows rapidly, pollution also increases rapidly. Rapid growth of the population also leads to other environmental issues such as the rapid depletion of natural resources.

The impacts of population on the quantitative water needs of a locality are related to population density (that is, how the population is distributed geographically), and to the rate of increase or decrease in population growth. Because population changes affect such variables as the economy, the environment, natural resources, the labor force, energy requirements, infrastructure needs, and food supply, they also affect the availability and quality of the water sources that can be drawn upon for use.

Food and Agriculture Organization of the United Nations

AQUASTAT
http://www.fao.org/nr/aquastat
Update: November 2016

Water withdrawal by sector, around 2010

Continent Regions	Subregions	Municipal km³/y	Municipal %	Industrial km³/y	Industrial %	Agricultural km³/y	Agricultural %	Total water withdrawal[①] km³/y	Total freshwater withdrawal km³/y	Freshwater withdrawal as % of IRWR
World		**464**	**12**	**768**	**19**	**2769**	**69**	**4001**	**3853**	**9**
Africa		33	15	9	4	184	81	227	220	6
Northern Africa		**14**	**13**	**3**	**3**	**89**	**84**	**106**	**101**	**215**
Sub-Saharan Africa		**19**	**16**	**6**	**5**	**96**	**79**	**121**	**119**	**3**
	Sudano Sahelian	2.1	5	0.6	1	40.2	94	42.8	42.8	26.8
	Gulf of Guinea	6.5	39	2.6	16	7.4	45	16.5	16.5	1.7
	Central Africa	1.3	45	0.5	19	1.0	36	2.8	2.8	0.1
	Eastern Africa	3.0	15	0.3	1	16.8	84	20.1	20.1	7.0
	Southern Africa	5.5	22	2.1	9	16.9	69	24.6	23.0	8.5
	Indian Ocean Islands	0.6	4	0.2	1	13.5	94	14.3	14.3	4.2
Americas		**123**	**14**	**321**	**37**	**415**	**48**	**859**	**855**	**4**
Northern America		**79**	**13**	**289**	**47**	**241**	**40**	**610**	**605**	**10**
	Northern America	68.0	13	281.5	53	179.8	34	529.3	526.0	9.3
	Mexico	11.4	14	7.3	9	61.6	77	80.3	79.5	19.4
Central America and Caribbean		**8**	**23**	**6**	**18**	**20**	**59**	**33**	**33**	**5**
	Central America	3.3	27	1.3	11	7.5	62	12.1	12.1	1.9
	Caribbean-Greater Antilles	4.0	19	4.6	22	12.0	58	20.5	20.5	22.2
	Caribbean-Lesser Antilles and Bahamas	0.4	60	0.1	23	0.1	18	0.6	0.5	9.7
Southern America		**36**	**17**	**26**	**12**	**154**	**71**	**216**	**216**	**2**
	Guyana	0.1	5	0.2	8	1.8	87	2.1	2.1	0.6
	Andean	10.9	18	3.9	7	45.2	75	60.1	60.0	1.1
	Brazil	17.2	23	12.7	17	44.9	60	74.8	74.8	1.3
	Southern America	7.9	10	9.0	11	62.4	79	79.3	79.1	5.7

Fig. 1

Food and Agriculture Organization of the United Nations

AQUASTAT
http://www.fao.org/nr/aquastat
Update: November 2016

Continent Regions	Subregions	Total withdrawal by sector						Total water withdrawal[①]	Total freshwater withdrawal	Freshwater withdrawal as % of IRWR
		Municipal		Industrial		Agricultural				
		km³/y	%	km³/y	%	km³/y	%	km³/y	km³/y	
Asia		234	9	253	10	2069	81	2556	2421	20
Middle East		25	9	20	7	231	84	276	268	55
	Arabian Peninsula	3.9	11	0.9	3	29.5	86	34.3	30.1	492.2
	Caucasus	1.7	10	2.9	17	12.3	73	16.9	16.9	23.1
	Islamic Republic of Iran	6.2	7	1.1	1	86.0	92	93.3	93.1	72.5
	Near East	13.6	10	14.9	11	103.3	78	131.8	128.0	46.3
Central Asia		7	5	10	7	128	89	145	136	56
Southern and Eastern Asia		202	9	224	10	1710	80	2135	2017	18
	South Asia	70.2	7	20.0	2	912.8	91	1003.1	889.6	46.0
	East Asia	98.3	14	158.0	22	469.4	65	725.8	721.5	21.2
	Mainland Southeast Asia	7.5	4	6.6	4	164.4	92	178.4	178.2	9.4
	Maritime Southeast Asia	25.7	11	39.1	17	163.4	72	228.2	228.0	5.9
Europe		69	21	181	54	84	25	334	332	5
Western and Central Europe		51	21	131	53	66	27	248	246	12
	Northern Europe	2.7	31	4.9	55	1.2	14	8.8	8.8	1.1
	Western Europe	21.0	21	73.5	74	4.9	5	99.4	98.7	15.9
	Central Europe	9.3	23	27.6	68	3.6	9	40.5	40.5	16.3
	Mediterranean Europe	17.9	18	25.0	25	55.9	57	98.8	97.8	23.1
Eastern Europe		18	21	50	58	18	21	86	86	2
	Eastern Europe	4.3	22	10.6	53	5.1	25	20.0	20.0	14.7
	Russian Federation	13.4	20	39.6	60	13.2	20	66.2	66.2	1.5
Oceania		5	20	4	15	16	65	25	25	3
Australia and New Zealand		5	20	4	15	16	65	25	24	3
Other Pacific Islands		0.03	30	0.01	11	0.05	59	0.1	0.1	0.1

① Includes use of desalinated water, direct use of treated municipal wastewater and direct use of agricultural drainage water.

Fig. 1 Water withdrawal by sector around 2010

2.1.2 Key points for social research

a. Population impacts on future water sources.

b. Population impacts on future water quality.

c. Reducing population impacts.

2.2 Water use for different purposes

2.2.1 Fundamentals

a. Municipal; b. Industrial; c. Domestic purposes.

2.2.2 Key points for social research

a. Types of agricultural water use: irrigation vs Rain-fed agriculture.

b. Types of industrial water use: cooling, cleaning/washing and employees' use.

c. Domestic water use.

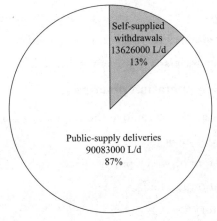

Fig. 2 Total domestic water use from public-supply deliveries and self-supplied withdrawals, for the United States 2010.

2.3 Water impurities

2.3.1 Fundamentals

The natural water contains solid, liquid, gaseous impurities and therefore, this water cannot be used for the generation of steam in the boilers. The impurities present in the water should be removed before it is used in steam generation. The necessity for reducing the corrosive nature & quantity of dissolved and suspended solids in feed water has become increasingly important with the advent of high pressure, critical & supercritical boilers.

Impurities in water: the impurities present in the feed water are classified as given below

a. Undissolved and suspended solid materials.

b. Dissolved salts and minerals.

c. Dissolved gases.

d. Other materials (as oil, acid) either in mixed or unmixed forms.

2.3.2 Key points for social research

a. Undissolved and suspended solid materials-turbidity and sediment; sodium and potassium salts; chlorides; iron; maganese; silica; microbiological growths; colour.

b. Dissolved salts and minerals-calcium and magnesium salts.

c. Dissolved gases-oxygen; carbon dioxide.

d. Other materials-free mineral acid; oil.

2.3.3 Outside operation practice

a. What is a way to remove the impurities from water without a filter or special equipment?

b. Parameters of water purity.

c. Purification methods.

d. Laboratory use.

e. Criticism strategy.

2.4 Waterborne diseases

2.4.1 Fundamentals

Waterborne diseases are caused by pathogenic microorganisms that most commonly are transmitted in contaminated fresh water. Infection commonly results during bathing, washing, drinking, in the preparation of food, or the consumption of food that is infected. Various forms of waterborne diarrheal diseases probably are the most prominent examples, and affect mainly children in developing countries; according to the World Health Organization, such diseases account for an estimated 3.6% of the total disability-adjusted life year global burden of diseases, and cause about 1.5 million human deaths annually. The World Health Organization estimates that 58% of that burden, or 842,000 deaths per year, is attributable to unsafe water supply, sanitation and hygiene.

2.4.2 Key points for social research

a. Socioeconomic impact of waterborne diseases.

b. Infections by types of pathogens (protozoa, bacteria, intestinal parasites and viruses).

2.4.3 Outside operation practice

a. Dimension of the problem caused by waterborne diseases.

b. Transmission of the waterborne diseases.

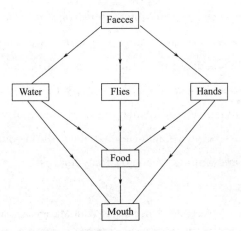

Fig. 3 Transmission of the waterborne diseases

c. Causes of waterborne diseases.

d. Preventing waterborne diseases.

2.5 Inorganic contaminants

2.5.1 Fundamentals

Inorganic contaminants, including heavy metals such as lead, cadmium and mercury, and other chemical elements such as arsenic and iodine, are elements or compounds found in water supplies and may be natural in the geology or caused by activities of man through mining, industry or agriculture. It is common to have trace amounts of many inorganic contaminants in water supplies. Amounts above the maximum contaminant levels may cause a variety of damaging effects to the liver, kidney, nervous system, circulatory system, blood, gastrointestinal system, bones, or skin depending upon the inorganic contaminant and level of exposure.

2.5.2 Key points for social research

If your water system exceeds any of the maximum contaminant levels, you must:

a. Notify the drinking water program and complete public notices as required.

b. Work with the drinking water program to determine the best way to reduce the level of contaminant in your water supply. A variety of options can be considered including a new treatment process, mixing your contaminated supply with another supply that does not exceed the maximum contaminant level, or using a new source of water.

c. Contact resource agencies listed on the resource agencies for help in planning and finding financing for your system improvements.

d. Monitor quarterly.

Table 1　Inorganic Contaminants With Maximum Contaminant Levels

Antimony	0.006mg/L	Fluoride	4mg/L
Arsenic	0.01mg/L	Mercury	0.002mg/L
Asbestos (measured by fibers)	7million/L	Nickel	0.1mg/L
Barium	2mg/L	Nitrate	10mg/L
Beryllium	0.004mg/L	Nitrite(as N)	1mg/L
Cadmium	0.005mg/L	Nitrate & Nitrite(combined)	10mg/L
Chromium	0.1mg/L	Selenium	0.05mg/L
Cyanide	0.2mg/L	Thallium	0.002mg/L

2.5.3 Outside operation practice

a. Pathways by which inorganic contaminants enter food.

b. Levels at which inorganic contaminants occur.

c. Mechanisms by which inorganic contaminants accumulate through the food chain.

2.6 Organic contaminants

2.6.1 Fundamentals

A long list of organic chemicals, including benzene, PCBs and vinyl chloride, can enter drinking water, potentially causing damage that includes increased cancer risk; liver, kidney, stomach, nervous system and immune system problems; reproductive difficulties; cataracts; and anemia. The water-contaminating chemicals come from farm runoff of herbicides or pesticides, leach from storage tanks and landfills and are discharged from factories. Many are known to be volatile organic compounds. Some continue on in the environment, as residue from pesticides that have long since been banned (like DDT) and others are added to the water during waste treatment (like chlorine).

2.6.2 Key points for social research

a. The effects of organic effluents on receiving waters-when an organic polluting load is discharged into a river, it is gradually eliminated by the activities of microorganisms in a way very similar to the processes in the sewage treatment works. This self-purification requires sufficient concentrations of oxygen, and involves the breakdown of complex organic molecules into simple in organic molecules. Dilution, sedimentation and sunlight also play a part in the process. Attached microorganisms in streams play a greater role than suspended organisms in self-purification. Their importance increases as the quality of the effluent increases since attached microorganisms are already present in the stream, whereas suspended ones are mainly supplied with the discharge.

b. Removing organic contaminants-activated carbon, ultrafiltration, reverse osmosis, organic scavenging filters, ozone injection, etc.

c. Common organic contaminants in wastewater-biochemical oxygen demand (BOD), chemical oxygen demand (COD) and total organic carbon (TOC).

2.6.3 Outside operation practice

a. Effects of organic contaminants on freshwater ecosystems.

b. Effects of organic contaminants on health.

c. What are the origins of organic contaminants?

2.7 Drinking water quality standards of China

2.7.1 Fundamentals

Drinking water quality standards describes the quality parameters set for drinking water. Despite the truism that every human on this planet needs drinking water to survive and that water may contain many harmful constituents, there are no universally recognized and accepted international standards for drinking water. Even where standards do exist, and are applied, the permitted concentration of individual constituents may vary by as much as ten times from one set of standards to another.

(1) The National Standards of the People's Republic of China-Environmental quality standards for surface water

This standard is hereby formulated for implementing the Environmental Protection Law and Law of Water Pollution Prevention and Control of People's Republic of China, and to control water pollution and to protect water resources.

This standard is applicable to the surface water bodies of rivers, lakes and reservoirs within the territory of the People's Republic of China.

(2) The classification of water bodies by functions

The water bodies are divided into five classes according to the utilization purposes and protection objectives:

Class I is mainly applicable to the water from sources, and the national nature reserves.

Class II is mainly applicable to first class of protected areas for centralized sources of drinking water, the protected areas for rare fishes, and the spawning fields of fishes and shrimps.

Class III is mainly applicable to second class of protected areas for centralized sources of drinking water, the protected areas for the common fishes and swimming areas.

Class IV is mainly applicable to the water areas for industrial use and entertainment which is not directly touched by human bodies.

Class V is mainly applicable to the water bodies for agricultural use and landscape requirement.

The water bodies with various functions are classified based on the highest function, and those with seasonal functions may be classified by seasons.

Table 2 The Environmental Quality Standards for Surface Water   mg/L

Parameters/ standards' value/ classification	I	II	III	IV	V
Fundamental requirements	All the water bodies should not contain the following substances resulted by non-natural reasons. a. Substances that can precipitate and become disgusting sediments; b. Floating matter, such as fragment, floating slag, oil and other matter that leads to unpleasant perception; c. Substances that produce disgusting color, smell and turbidity; d. Substances that are harmful, toxic to or has negative physical effects on human beings, animals and plants; e. Substances that benefits the disgusting aquatic organism					
Water temperature/℃	The water temperature variations caused by human activities should be controlled within: The maximum average weekly temperature rise in summer$\leqslant 1$; The maximum average weekly temperature drop$\leqslant 2$					

Continued Table 2

Parameters/ standards' value/ classification	I	II	III	IV	V
pH	6.5~8.5	6.5~8.5	6.5~8.5	6.5~8.5	6~9
Sulfate[①] (by SO_4^{2-})	Lower than 250	250	250	250	250
Chloride[①] (by Cl^-)≤	Lower than 250	250	250	250	250
Soluble Fe[①]≤	Lower than 0.3	0.3	0.5	0.5	1.0
Total Manganese[①] (Mn)≤	Lower than 0.1	0.1	0.1	0.5	1.0
Total Copper[①]≤	Lower than 0.01	1.0 (fishery 0.01)	1.0 (fishery 0.01)	1.0	1.0
Total Zinc[①]≤	0.05	1.0 (fishery 0.1)	1.0 (fishery 0.1)	2.0	2.0
Nitrate(by N)≤	Lower than 10	10	20	20	25
Nitrite(by N)≤	0.06	0.1	0.15	1.0	1.0
Non-ionic Nitrogen≤	0.02	0.02	0.02	0.2	0.2
Kjeldahl Nitrogen≤	0.5	0.5	1	2	2
Total Phosphorus (by P)≤	0.02	0.1(lakes, reservoirs 0.025)	0.1(lakes, reservoirs 0.05)	0.2	0.2
Permanganate Index≤	2	4	6	8	10
Soluble Oxygen≤	Percentage of saturation 90%	6	5	3	2
Chemical Oxygen Demand(COD_{Cr})≤	Lower than 15	Lower than 15	15	20	25
Biochemical Oxygen Demand(BOD_5)≤	Lower than 3	3	4	6	10
Fluoride(F^-)≤	Lower than 1.0	1.0	1.0	1.5	1.5
Selenium(four)≤	Lower than 0.01	0.01	0.01	0.02	0.02
Total Arsenic≤	0.05	0.05	0.05	0.1	0.1
Total Mercury[②]≤	0.00005	0.00005	0.0001	0.001	0.001
Total Cadmium[③]≤	0.001	0.005	0.005	0.005	0.01

Continued Table 2

Parameters/ standards' value/ classification	I	II	III	IV	V
Total Chromium (six)⩽	0.01	0.05	0.05	0.05	0.1
Total Lead[2]⩽	0.01	0.05	0.05	0.05	0.1
Total Cyanide⩽	0.005	0.05 (fishery 0.005)	0.2 (fishery 0.005)	0.2	0.2
Volatile Phenol[2]⩽	0.002	0.002	0.005	0.01	0.1
Oil Category[2]⩽	0.05	0.05	0.05	0.5	1.0
Anionic Surface-active Agent⩽	Lower than 0.2	0.2	0.2	0.3	0.3
Total Coliform Group Bacteria[3]/L⩽			1000		
Benzo(a) pyrene[3]/(μg/L)⩽	0.0025	0.0025	0.0025		

① The items permitted to be appropriately adjusted according to the background characteristic of the local water body.

② The lowest measurement values in the stipulated analysis and measurement methods, which can not meet the fundamental requirements.

③ Tentative standards.

Table 3 Drinking water quality standards of China

Index	Maximum level
(1) Microbial index	
Total coliforms/(MPN/100mL or CFU/100mL)	Negative
Thermotolerant coliforms/(MPN/100mL or CFU/100mL)	Negative
Escherichia coli (E. coli)/(MPN/100mL or CFU/100mL)	Negative
Colony count/(CFU/mL)	100
(2) Toxicological index	
Arsenic/(mg/L)	0.01
Cadmium/(mg/L)	0.005
Chromium(Ⅵ)/(mg/L)	0.05

Continued Table 3

Index	Maximum level
Lead/(mg/L)	0.01
Mercury/(mg/L)	0.001
Selenium/(mg/L)	0.01
Cyanide/(mg/L)	0.05
Fluoride/(mg/L)	1.0
Nitrates(indicated by N)/(mg/L)	10(20 for groundwater)
Chloroform/(mg/L)	0.06
Carbon tetrachloride(CCl_4)/(mg/L)	0.002
Bromate(when using ozone)/(mg/L)	0.01
Formaldehyde(when using ozone)/(mg/L)	0.9
Chlorite (when using chlorine dioxide to disinfect)/(mg/L)	0.7
Chlorate(when using compound chlorine dioxide to disinfect)/(mg/L)	0.7
(3)Sensory properties and general chemical indicator	
Colour(Platinum cobalt unit)	15
Turbidity(NTU-scattering turbidity unit)	1 (3 for water source and water purification technical conditions)
Smell and taste	No foul and smell
Visible substances	No
pH	6.5~8.5
Aluminum/(mg/L)	0.2
Iron/(mg/L)	0.3
Manganese/(mg/L)	0.1
Copper/(mg/L)	1.0
Zinc/(mg/L)	1.0
Chloride/(mg/L)	250
Sulfate/(mg/L)	250
Total dissolved solids/(mg/L)	1000
Total hardness(in $CaCO_3$)/(mg/L)	450
Oxygen consumption(COD_{Mn} method, in O_2)/(mg/L)	3 (5 for water source/when the oxygen consumption of raw water exceeds 6mg/L)

Continued Table 3

Index	Maximum level
Volatile phenols(in phenol)/(mg/L)	0.002
Anionic synthetic detergent/(mg/L)	0.3
(4)Radioactivity index	Guideline value
Total alpha(α)radioactivity/(Bq/L)	0.5
Total beta(β)radioactivity/(Bq/L)	1
MPN represents the most probable number; CFU stands for colony forming units. When the total coliform group is detected in water samples, escherichia coli or thermotolerant coliform bacteria should be further examined. If the total coliform bacteria are not detected in the water samples, there is no need to test E. *coli* or thermotolerant coliform bacteria	If the radioactivity index exceeds the guiding value, radionuclide analysis and evaluation should be carried out to determine whether or not it can be drunk

2.7.2 Key points for social research

a. Water quality feasibility test and water environment management system.

b. Sustainable plan for China's drinking water.

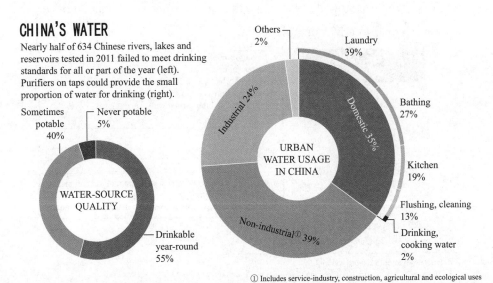

Fig. 4 Sustainable plan for China's drinking water

2.7.3 Outside operation practice

a. What is the purpose of drinking water quality guidelines/regulations?

b. What is the quality of Chinese drinking water?

c. The importance of clean drinking water.

2.8 Drinking water quality standards of Japan

2.8.1 Fundamentals

In July 2002, in response to the undergoing discussion of the 3rd Guidelines for WHO's Drinking Water Quality Guidelines, the ministry consulted with the Health Science Council on Japan's basic concept of Drinking Water Quality Standards revision.

A special committee for water quality was formed to discuss the concept of the revision of standards as well as the water quality management systems such as examination frequency. In April 2003, according to the report submitted by the council and the result of public comments, the new standards and the water quality management systems were formed and went into effect in April 2004.

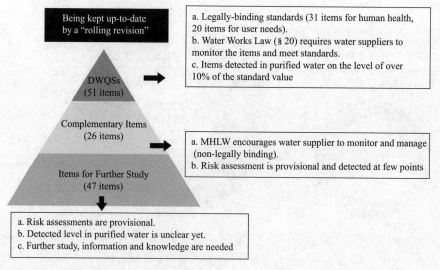

Fig. 5 Scheme of drinking water quality control

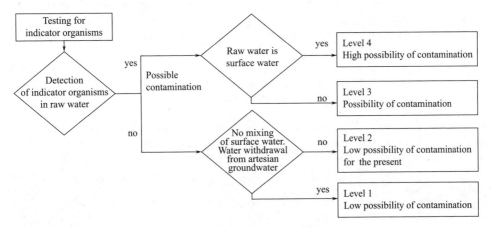

Fig. 6 Determination of risk level

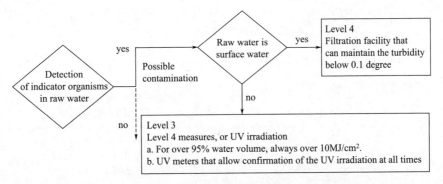

Fig. 7 Required facilities

Ministry of Health, Labour and Welfare in Japan is established the drinking water quality standards of 46 items to keep up the supply of drinking water with clean and safe. To respond to the standards, all the water utilities have fully introduced the works of improvement of water treatment facilities as well as introduction of necessity operation management. On the other hand, the water quality laboratory of water utilities is periodically conducting in order to measure for water quality whether tap water meets the standards perfectly by these measures or not.

Table 4 Water quality standards of drinking water in Japan

No	Item	Standard Value	No	Item	Standard Value
1	Standard Plate Count	100 in 1mL or less	28	Bromodichloromethane	0.03mg/L or less
2	Coliform Group Bacterial	Not to be detected	29	Bromoform	0.09mg/L or less
3	Cadmium	0.01mg/L or less	30	Formaldehyde	0.08mg/L or less
4	Mercury	0.0005mg/L or less	31	Zinc	1.0mg/L or less
5	Selenium	0.01mg/L or less	32	Aluminium	0.2mg/L or less
6	Lead	0.01mg/L or less	33	Iron	0.3mg/L or less
7	Arsenic	0.01mg/L or less	34	Copper	1.0mg/L or less
8	Chromium(Ⅵ)	0.05mg/L or less	35	Sodium	200mg/L or less
9	Cyanide	0.01mg/L or less	36	Manganese	0.05mg/L or less
10	Nitrates/Nitrites	10mg/L as nitrogen or less	37	Chloride	200mg/L or less
11	Fluoide	0.8mg/L or less	38	Calcium Magnesium, etc. (Hardness)	300mg/L or less
12	Boron	1.0mg/L or less	39	Total Residue	500mg/L or less
13	Carbon Tetrachloride	0.002mg/L or less	40	Methylene Blue Activated Substance	0.2mg/L or less
14	1,4-dioxane	0.05mg/L or less			
15	1,1-dichloroethylene	0.02mg/L or less	41	Geosmin	0.00001mg/L or less[①]
16	cis-1,2-dichloroethylene	0.04mg/L or less	42	2-methylisoborneol (MIB)	0.00001mg/L or less[①]
17	Dichloromethane	0.02mg/L or less	43	Nonionic Surfactant	0.02mg/L or less
18	Tetrachloroethylene	0.01mg/L or less	44	Phenols	0.005mg/L as phenol or less[②]
19	Trichloroethylene	0.03mg/L or less			
20	Benzene	0.01mg/L or less	45	Organic Compound (as concentration of TOC)	5mg/L or less
21	Chloroacetic Acid	0.02mg/L or less			
22	Chloroform	0.06mg/L or less			
23	Dichloroacetic Acid	0.04mg/L or less	46	pH Value	5.8-8.6
24	Dibromochloromethane	0.1mg/L or less	47	Taste	Not abnormal
25	Bromate	0.01mg/L or less	48	Odor	Not abnormal
26	Total Trihalomethanes	0.1mg/L or less	49	Color	5 degree or less
27	Trichloroacetic Acid	0.02mg/L or less	50	Turbidity	2 degree or less

① The standard value is 0.00002mg/L by 31 March 2007.

② The standard value of organic compound etc. as the potassium permanganate consumption is 10mg/L by 31 March 2005.

2.8.2 Key points for social research

Introduction of Water Safety Plan is expected to result in effects such as the following:

a. Unified identification and evaluation of water supply systems.

b. Reduction of risks and improvements in safety
-Re-evaluation of systems by means of objective methods;
-Elimination of unfounded assumptions.

c. Improvement and streamlining of maintenance and control levels
-Clarification of priority regarding crucial control points;
-Improvement of operators' maintenance and control ability.

d. Transfer of technology to the next generation (through unified documentation).

e. Improvement of communications with concerned third parties.

f. Increase of accountability on water safety to consumers.

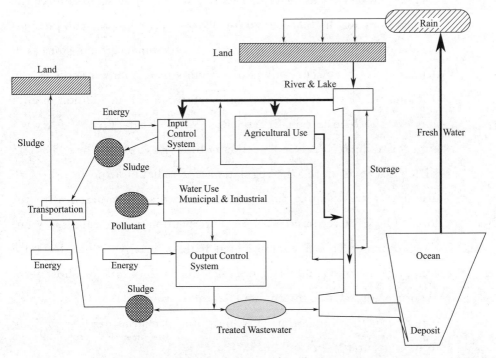

Fig. 8 Scheme of water cycle and water uses

g. Approach to watershed areas-related stakeholders in order to secure better water quality at water sources.

2.8.3 Outside operation practice

a. The state of the water quality in Japan.

b. Infectious diseases caused by drinking water.

c. The present state of water resources in Japan.

d. Water cycle and water uses.

2.9 US EPA (Environmental Protection Agency) water quality standards

2.9.1 Fundamentals

EPA ensures safe drinking water for the public, by setting standards for more than 160000 public water systems nationwide. EPA oversees states, local governments and water suppliers to enforce the standards under the *Safe Drinking Water Act*. The program includes regulation of injection wells in order to protect underground sources of drinking water. Select readings of amounts of certain contaminants in drinking water, precipitation, and surface water, in addition to milk and air, are reported on EPA's Rad Net web site (http://www.epa.gov/radnet/EPA) in a section entitled Envirofacts. In 2013, an EPA draft revision relaxed regulations for radiation exposure through drinking water, stating that current standards are impractical to enforce. The EPA recommended that intervention was not necessary until drinking water was contaminated with radioactive iodine 131 at a concentration of 81000 picocuries per liter (the limit for short term exposure set by the International Atomic Energy Agency), which was 27000 times the prior EPA limit of 3 picocuries per liter for long term exposure.

Table 5 Drinking water standards and health advisories-water quality, secondary standards

Component	Standard	EPA	IRIS		Health Advisories			Wt. of Evid.	California		Arizona	Hawaii
		SMCL	RfD μg/(kg·d)	10^{-6} Risk	Acute 10 days	Chronic(lifetime)			MCL	Action Level	MCL	MCL
						Non-Cancer	Cancer					
Color	Secondary	15 color units							①			
Corrosivity	Secondary	Non-corrosive							①			
Foaming Agents	Secondary	500μg/L							①			
Odor	Secondary	3.0 OT②							①			
Total dissolved solids(TDS)	Secondary	500mg/L							①			
pH	Secondary	6.5-8.5										

① Secondary standards are not federally enforceable, but may be enforced in California.
② Odor threshold number.

Table 6　Drinking water standards and health advisories-microbials and indicators

Contaminant	Standard	EPA		IRIS			Health Advisories			Wt. of Evid.	California		Arizona	Hawaii
		MCL	MCLG	RfD μg/(kg·d)	10^{-6} Risk		Acute 10 days	Chronic(lifetime)			MCL	Action Level	MCL	MCL
								Non-Cancer	Cancer					
Cryptosporidium	Current	TT a[①]	0											
Giardia lamblia	Current	TT b[②]	0											
Heterotrophic Plate Count	Current	TT c[③]												
Legionella	Current	TT c	0											
Total Coliform Bacteria	Current	P/A[⑤]	0											
Turbidity	Current	0.3/1 NTU[⑥]												
Viruses	Current	TT d[④]	0											

① Two-log reduction by filtration treatment technique applies to surface water systems serving > 10000 people.
② Three-log filtration/inactivation treatment technique applies to all surface water systems.
③ Applies to surface water systems only.
④ Four-log inactivation treatment technique applies to all surface water systems, groundwater systems that require disinfection.
⑤ MCL is presence/absence of total or fecal coliform bacteria.
⑥ 0.3 NTU, conv. or direct filtration; 1 NTU, diatomaceous earth or slow sand filtration. Applies only to systems required to filter.

Table 7 Drinking water standards and health advisories-inorganic (1)

Chemical	Standard	EPA		IRIS		Health Advisories			Wt. of Evid.	California[1]			Arizona[1]	Hawaii[1]
						Acute	Chronic(lifetime)							
		MCL	MCLG	RfD $\mu g/(kg \cdot d)$	10^{-6} Risk	10 days	Non-Cancer	Cancer		MCL	PHG		MCL	MCL
Aluminum	Secondary	50-200	0							1000 200 SMCL	600			
Ammonia							30000		D					
Antimony	Current	6	6	0.4		15	3		D	6	20		6	6
Arsenic	Current	10	0	0.3	0.02			0.02	A	50			50	50
Asbestos	Current	7E+6 10μm fibers	7E+6 10μm fibers						A	7E+6 10μm fibers				
Barium	Current	2000	2000	200		700			N	1000	700		2000	2000
Beryllium	Current	4	4	2		30000				4	1		4	4
Boron				200		900	1000		I					
Bromate	Current	10	0	4	0.05			0.05	B2					
Cadmium	Current	5	5	0.5		40	5		D	5	0.07		5	5

Chapter 2 Water use and water quality

续表

Chemical	Standard	EPA		IRIS		Health Advisories			Wt. of Evid.	California[①]		Arizona[①]	Hawaii[①]
		MCL	MCLG	RfD μg/(kg·d)	10^{-6} Risk	Acute 10 days	Chronic(lifetime)			MCL	PHG	MCL	MCL
							Non-Cancer	Cancer					
Chloramines	Current	MRDL[②] 4.0mg/L as Cl	MRDLG[②] 4.0mg/L as Cl	100		1000	3000		D				
Chlorate									D		200		
Chloride	Secondary	250mg/L								250–600 Secondary			
Chlorine	Current	MRDL[②] 4.0mg/L as Cl	MRDLG[②] 4.0mg/L as Cl	100		3000	4000		D				

Values are indicated in micrograms per liter (μg/L)[equivalent to parts per billion (ppb)]unless otherwise stated.

Oral Referenced Doses (RfD) are in micrograms per kilogram per day(μg/kg·d), 10^{-6} lifetime risk levels are in micrograms per liter.

① EPA MCLs apply unless noted as different.

② MRDL,MRDLG:Maximum residual disinfectant level and goal. Apply only if this disinfectant is used.

Table 8 Drinking water standards and health advisories-inorganic (2)

Chemicals	Standard	EPA			IRIS			Health Advisories			Wt. of Evid.	California[1]			Arizona[1]	Hawaii[1]
		MCL	MCLG	MRDL[2] / MRDLG[2]	RfD μg/(kg·d)	10^{-6} Risk	Acute 10 days	Chronic(lifetime) Non-Cancer	Chronic(lifetime) Cancer			MCL	PHG		MCL	MCL
Chlorine Dioxide	Current			0.8mg/L as ClO2 / 0.8mg/L as ClO2	30		840	800		D					
Chlorite	Current	1.0mg/L	800		30		840	800		D					
Chromium (total)	Current	100	100		3		1000.00			D	50			100	100
Copper	Current / Secondary	AL 1300 TT##[5] 1000	1300							D	TT##[6] 1000 SMCL	170		TT##[6]	TT##[6]
Cyanide	Current	200	200		22		200	200		D	200	150		200	200
Fluoride	Current / Proposed / Secondary	4mg/L 2mg/L	4mg/L		60						1400~ 2400td[4]	1000		4mg/L	4mg/L
Iron	Secondary	300									300				
Lead	Current	AL 15 TT#[5]	0							B2	TT#[5]	2		TT#[5]	TT#[5]

Chapter 2　Water use and water quality　27

续表

Chemicals	Standard	EPA			IRIS		Health Advisories			Wt. of Evid.	California①		Arizona①	Hawaii①
		MCL	MCLG		RfD $\mu g/(kg \cdot d)$	10^{-6} Risk	Acute 10 days	Chronic(lifetime) Non-Cancer	Chronic(lifetime) Cancer		MCL	PHG	MCL	MCL
Manganese	Secondary	50			140(food) 5(water)						50			
Mercury (inorganic)	Current	2	2		0.3			2		D	2	1.2	2	2
Molybdenum					5		40	40		D				
Nickel					20		1000	100		D	100	12		
Nitrate (as N)	Current	10mg/L	10mg/L		1.6mg/L		10 mg/L④			D	45mg/L (as NO_3^-)		10mg/L	10mg/L

Value are indicated in micrograms per liter ($\mu g/L$) [equivalent to parts per billion (ppb)] unless otherwise stated.
Oral Reference Doses (RfD) are in micrograms per kilogram per day [$\mu g/(kg \cdot d)$], 10^{-6} lifetime risk levels are in micrograms per liter.

① EPA MCLs apply unless noted as different.
② MRDL, MRDLG: Maximum residual disinfectant level and goal. Apply only if this disinfectant is used.
③ 10 day NA for nitrate/nitrite for 4kg child (protective of 10kg child & adults); also used for chronic(lifetime).
④ temperature dependent value.
⑤ Treatment technique and public notification triggered at Action Level of 15 ppb.
⑥ Treatment technique triggered at Action Level of 1300 ppb.

2.9.2 Key points for social research

National Guidance on the Applicability of Freshwater and Saltwater Criteria

EPA recommends that the aquatic life criteria in this compilation apply as follows:

a. For water in which the salinity is equal to or less than 1 part per thousand 95% or more of the time, the applicable criteria are the freshwater criteria.

b. For water in which the salinity is equal to or greater than 10 parts per thousand 95% or more of the time, the applicable criteria are the saltwater criteria in Column C.

c. For water in which the salinity is between 1 and 10 parts per thousand, the applicable criteria are the more stringent of the freshwater or saltwater criteria. However, an alternative freshwater or saltwater criteria may be used if scientifically defensible information and data demonstrate that on a site-specific basis the biology of the water body is dominated by freshwater aquatic life and that freshwater criteria are more appropriate; or conversely, the biology of the water body is dominated by saltwater aquatic life and that saltwater criteria are more appropriate.

2.9.3 Outside operation practice

a. What is the relationship between these criteria and your State or tribal water quality standards?

b. Where can I find more information about water quality criteria and water quality standards?

c. What are the national recommended water quality criteria?

Chapter 3 Water sources and water intake facilities

When choosing a technology, the rationale for using a particular water source should be considered. Several types of water sources, such as wells, ponds, rivers or springs are traditionally used for different purposes and they may not be operational all year. Some water sources are more reliable, convenient, or provide water that tastes better. If users percieve an "improvement" as something "worse" in any one aspect, they may return to their traditional, contaminated source. Chlorinating water, for instance, may introduce odour or taste and it may be necessary to explain the need for chlorination to users.

The following list of community water sources and intake technologies is not exhaustive, but represents those most commonly found in developing countries:

(1) **Rainwater**

a. Rooftop rainwater harvesting;

b. Catchment and storage dams.

(2) **Surface water**

a. Protected side intake;

b. River-bottom intake;

c. Floating intake;

d. Sump intake.

(3) **Groundwater**

a. Springwater collection;

b. Dug well;

c. Drilled wells;

    d. Subsurface harvesting systems.

3.1 Rainwater

3.1.1 Rooftop rainwater harvesting

Rooftop catchment (Fig. 9) systems gather rainwater from the roofs of houses, schools, etc., using gutters and downpipes (made of local wood, bamboo, galvanized iron or PVC), and lead it to storage contain-ers that range from simple pots to large ferrocement tanks. If properly designed, a foul-flush device or detachable down-pipe can be fitted that allows the first 20 litres of runoff from a storm to be diverted from the storage tanks. The run off is generally contaminated with dust, leaves, insects and bird droppings. Sometimes the runoff is led through a small filter of gravel, sand and charcoal before entering the storage tank. Water may be abstracted from the storage tank by a tap, hand-pump, or bucket-and-rope system.

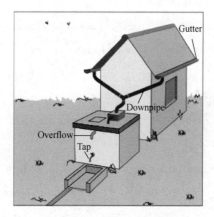

Fig. 9   Rooftop catchment

(1) Key points for operation

Operations consist of taking water from the storage tank by tapping, pumping or using a bucket and rope. Where there is no foul-flush device, the user or caretaker has to divert away the first 20 litres at the start of every rainstorm, and keep the rooftop clean. Just before the start of the rainy season, the complete system has to be checked for holes and for broken or affected parts, and repaired as necessary. Taps or handpumps (if used) have to be serviced. During the rainy season, the system should be checked regularly and cleaned when dirty. The system should be also checked and cleaned after every dry period of more than one month. The outsides of metal tanks may need to be painted about once a year. Leaks have to be repaired throughout the year, especially from leaking tanks and taps, as they present health risks. Chlorination of the water may be necessary.

(2) Outside operation practice

a. Monitoring the condition of the system and the water quality;

b. Providing access to credit facilities for buying or replacing a system;

c. Training users/caretakers for management;

d. Training local craftsmen to carry out larger repairs.

(3) Potential problems

a. Corrosion of metal roofs, gutters, etc. ;

b. The foul-flush diverter fails because maintenance was neglected;

c. Taps leak at the reservoir and there are problems with the handpumps;

d. Contamination of uncovered tanks, especially where water is abstracted with a rope and bucket;

e. Unprotected tanks may provide a breeding place for mosquitoes, which may increase the danger of vector-borne diseases;

f. During certain periods of the year, the system may not fulfil drinking-water needs, making it necessary to develop other sources, or to go back to traditional sources during these periods;

g. Households or communities cannot afford the financial investment needed to construct a suitable tank and adequate roofing.

3.1.2 Catchment and storage dams

Water can be made available by damming a natural rainwater catchment area, such as a valley, and storing the water in the reservoir formed by the dams, or diverting it to another reservoir. Important parameters in the planning of dams are: the annual rainfall and evaporation pattern; the present use and runoff coefficient of the catchment area; water demand; and the geology and geography of the catchment area and building site. Dams can consist of raised banks of compacted earth (usually with an impermeable clay core, stone aprons and a spillway to discharge excess runoff), or masonry or concrete (reinforced or not). In this manual, we refer only to dams less than a few metres high. The water stored behind a dam should normally be treated before entering a distribution system (Fig. 10).

Fig. 10　A small dam

(1) Key points for operation

Caretaker operations can include activities such as opening or closing valves or sluices in the dams, or in conduits to the reservoirs. The actual water collection from water points is usually carried out by the users themselves, often women and children.

If the water is for human consumption, cattle and people should be kept away from the catchment area and reservoir all year round. This can be helped by having a watch-man patrol regularly, and by fencing off the area. Water should be provided to users through a treatment plant and a distribution system with public standpipes or household connections. The dams, valves, sluices and pipelines have to be checked for leaks and structural failures. If repairs cannot be carried out immediately, the points of failure should be marked. The catchment area must also be checked for contamination and erosion. To control erosion, grass or trees could be planted just before the rainy season, and a nursery may have to be started. Once a year, the reservoirs may be left to dry out for a short period to reduce the danger of bilharzia. The reservoirs, silt traps, gutters, etc., must be de-silted at least once a year. To control mosquito breeding and the spread of malaria, Tilapia fish can be introduced in the reservoirs (every year if it runs dry).

(2) Outside operation practice

a. The water consumption allowed for each user;

b. Preventing water contamination;

c. Solving upstream-downstream conflicts.

3.2 Surface water

3.2.1 Fundamentals

Surface water is water on the surface of the planet such as in a river, lake, wetland or ocean. It can be contrasted with groundwater and atmospheric water. Non-saline surface water is replenished by precipitation and by recruitment from ground-water. It is lost through evaporation, seepage into the ground where it becomes ground-water, used by plants for transpiration, extracted by mankind for agriculture, living, industry, etc, or discharged to the sea where it becomes saline.

(1) Water storage in surface water

Rain (or precipitation in general) can be stored in surface water like rivers, lakes and ponds. The water level will rise. Later, the water will either be transported downstream, seep into the soil (infiltration) or evaporate. Then the level will be normal again. The amount of water that can be stored depends on the size of the basin and the height that it can rise before it overflows. When there is more rain than can be stored, infiltrated, or transported the area will flood. The surface water is very important for the temporary storage of water. Drainage or discharge of water out of the area is always a slower process, than storage close to where the water falls. The process of water storage in surface water is shown in Fig. 11.

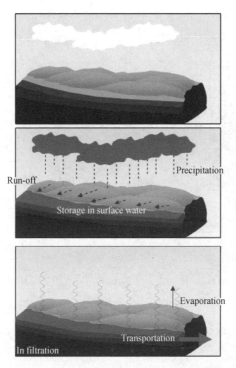

Fig. 11 The process of water storage in surface water

(2) Water storage in soil

Rain water can also be stored in the ground (Fig. 12). Soils consist of particles and pores. Those pores can be filled not only with air but also with water. The amount of pores is a soil is different for different types of soil. The pores in a clay soil account for 40%-60% of the volume. In fine sand this can be 20%-45%. The soil particles have small pores in them where water can enter (soil water) and between the particles are larger pores that can be filled. The soil is filled with water up a certain level. This level goes up and down with changing weather conditions. This water level is the ground water level. The process of water entering the soil is called infiltration. When the soil has taken up all the water it can, we say that it is saturated. If you walk over a saturated soil, you feel that it is wet and soggy, like biscuits dipped in tea. Part of the water that infiltrates, will move on. It will go to underground storage reservoirs or to underground rivers and may, through ground water flows, eventually reach a river or a lake. Another part will be used by plants or will evaporate.

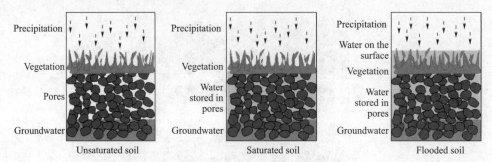

Fig. 12　The process of water storage in soil

(3) Water storage in sewage systems

In areas with a lot of hard surfaces, like cities with streets and houses, the water runs off immediately to sewers or becomes surface water. As there is hardly any water storage in the soil in areas with hard surfaces, you see a quick rising of the surface water level when it rains. When the sewers are used to capacity the streets will flood. The process of water storage in sewage systems is shown in Fig. 13.

3.2.2　Key points for social research

  a. Surface water treatment rules;

  b. Surface water resources in major river basins of China;

  c. Surface water protection programs.

3.2.3　Outside operation practice

  a. Summary of surface-water use in China, Japan, U.S., etc.

  b. Surface water information;

Fig. 13　The process of water storage in sewage systems

　　c. Factors affecting surface runoff;

　　d. Which type of drainage basin has the greatest effect on surface runoff?

3.3　Groundwater

3.3.1　Fundamentals

　　Groundwater is the water present beneath earth's surface in soil pore spaces and in the fractures of rock formations. A unit of rock or an unconsolidated deposit is called an aquifer when it can yield a usable quantity of water. The depth at which soil pore spaces or fractures and voids in rock become completely saturated with water is called the water table. Groundwater is recharged from, and eventually flows to, the surface naturally; natural discharge often occurs at springs and seeps, and can form oases or wetlands. Groundwater is also often withdrawn for agricultural, municipal and industrial use by constructing and operating extraction wells. Groundwater flow is shown in Fig. 14.

　　Why is there groundwater?

　　A couple of important factors are responsible for the existence of groundwater:

　　a. Gravity. Nothing surprising here: gravity pulls water toward the

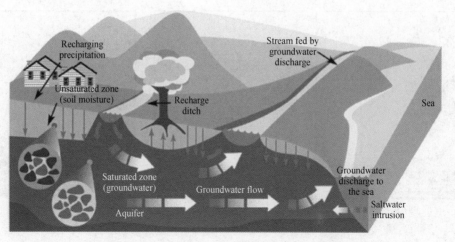

Fig. 14　Groundwater flow

center of the Earth. That means that water on the surface will try to seep into the ground below it.

b. The rocks below our feet. The rock below the Earth's surface is the bedrock. If all bedrock consist of a dense material like solid granite, then even gravity would have a hard time pulling water downward. But Earth's bedrock consists of many types of rock, such as sandstone, granite, and limestone. Bedrock have varying amounts of void spaces in them where groundwater accumulates. Bedrock can also become broken and fractured, creating spaces that can fill with water. And some bedrock, such as limestone, are dissolved by water—which results in large cavities that fill with water.

3.3.2　Key points for social research

a. Groundwater—a major link in the hydrologic cycle;

b. Groundwater—always on the move;

c. Safeguarding our groundwater supply;

d. Groundwater quality;

e. Groundwater and geology;

f. Groundwater and engineering;

g. Groundwater and wetlands;

h. Groundwater and permafrost;

i. Groundwater use (Groundwater as a source of energy; Almost nine million Canadians depend on groundwater).

3.3.3 Outside operation practice

a. How much groundwater do we have? (Fig. 15)

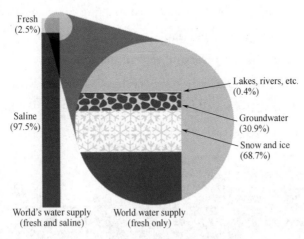

Fig. 15　Groundwater and the world's freshwater supply

Source: Adapted from Figure 2, Freshwater Series No. A-2, Water-Here, There and Everywhere.

b. Groundwater recharge. (Fig. 16)

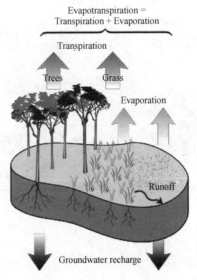

Fig. 16　Water balance

Chapter 3　Water sources and water intake facilities

3.4 Reservoir storage

3.4.1 Fundamentals

A reservoir usually means an enlarged natural or artificial lake, storage pond or impoundment created using a dam or lock to store water. Reservoirs can be created by controlling a stream that drains an existing body of water. They can also be constructed in river valleys using a dam. Alternately, a reservoir can be built by excavating flat ground or constructing retaining walls and levees. Tank reservoirs store liquids in storage tanks that may be elevated, at grade level, or buried. Tank reservoirs for water are also called cisterns. (Fig. 17)

Fig. 17 Types of reservoirs

a. Reservoirs dammed in valleys;
b. Bank-side reservoir;
c. Service reservoir.

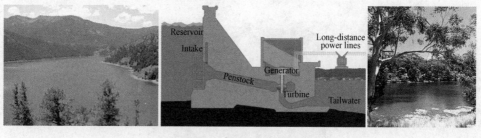

Fig. 18 Reservoirs use

a. Direct water supply;
b. Hydroelectricity;
c. Controlling watercourses;

    d. Flow balancing;

    e. Recreation: many reservoirs often allow some recreational uses, such as fishing and boating (Fig. 18).

3.4.2　Key points for operation

    a. Terminology: the terminology for reservoirs varies from country to country. In most of the world, reservoir areas are expressed in square kilometres; in the United States acres are commonly used. For volume either cubic metres or cubic kilometres are widely used, with acre-feet used in the United States.

    b. Modelling reservoir management.

3.4.3　Outside operation practice

    a. Whole life environmental impact of reservoirs.

    b. Climate change-Reservoir greenhouse gas emissions; hydroelectricity and climate change.

    c. Biology-Dams can produce a block for migrating fish, trapping them in one area, producing food and a habitat for various waterbirds. They can also flood various ecosystems on land and may cause extinctions.

    d. Human impact-Dams can severely reduce the amount of water reaching countries downstream of them, causing water stress between the countries, e.g. the Sudan and Egypt, which damages farming businesses in the downstream countries, and reduces drinking water.

3.5　Water intake facilities

3.5.1　Fundamentals

    Water intake facilities can be located in riverine, estuarine, and marine environments and can include domestic water supply facilities, irrigation systems for agriculture, power plants, and industrial process users. Nearly half of U.S. water withdrawals are attributed to thermoelectric power facilities, and about one-third are used for agriculture irrigation. In freshwater riverine systems, water withdrawal for commercial and domestic water use supports the needs of homes, farms, and industries that require a constant supply of water. Freshwater is diverted directly from lakes, streams, and rivers by means of pumping facilities or is stored in impoundments or reservoirs. Water withdrawn from estuarine and marine environments may be used to cool coastal power generating stations, as a source of water for agricultural purposes, and more recently, as a source of domestic water through desalinization facilities. In the case of power plants and desalinization plants, the subsequent discharge of water with temperatures higher than ambient levels can also occur.

(1) Springwater collection

The main parts of a springwater collection system are: a drain under the lowest natural water level; a protective structure at the source, for stability; a seal to prevent surface water from leaking back into the stored water (Fig. 19).

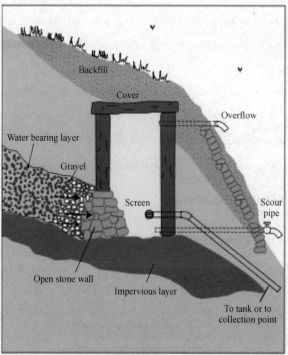

Fig. 19　Springwater collection

(2) Dug wells

These are wells that are dug by hand or by machinery, and consist of the following main parts: a stone, brick or concrete apron; a headwall (the part of the well lining above ground) at a convenient height for collecting water; a lining that prevents the well from collapsing (Fig. 20).

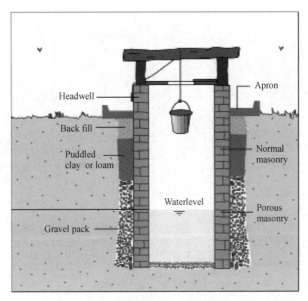

Fig. 20 Dug well

(3) Drilled wells

Boreholes can be constructed by machine or by hand-operated equipment, and usually consist of three main parts:

a. A concrete apron around the borehole at ground level (with an outlet adapted to the water abstraction method). This prevents surface water from seeping down the sides of the well, provides a hard standing, and directs lost water away from the well to a drainage channel.

b. A lining below the ground, but not going into the aquifer, to prevent it from collapsing, especially in unconsolidated formations. The lining is usually pipe material (mostly PVC and sometimes galvanized iron). In consolidated formations, the lining may not be required.

c. A slotted pipe below water level, to allow groundwater to enter the well. A layer of gravel surrounding the slotted pipe facilitates groundwater movement towards the slotted pipes and prevents ground material from entering the well. In consolidated formations, this gravel may not be required (Fig. 21).

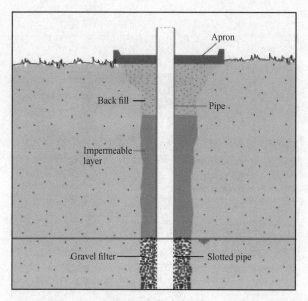

Fig. 21　Drilled well

(4) Subsurface harvesting systems

Subsurface harvesting systems retain groundwater flows and facilitate their abstraction. There are two main systems: subsurface dams (Fig. 22) and raised-sand dams.

(5) Protected side intake

A protected side intake provides a stable place in the bank of a river or lake (Fig. 23), from where water can flow into a channel or enter the suction pipe of a pump. It is built to withstand damage by floods and to minimize problems caused by sediment. Side intakes are sturdy structures, usually made of reinforced concrete, and may have valves or sluices to flush any sediment that might settle. Often, a protected side intake is combined with a weir in the river to keep the water at the re-

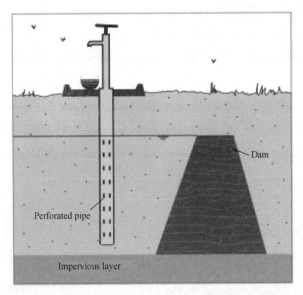

Fig. 22　Subsurface dam

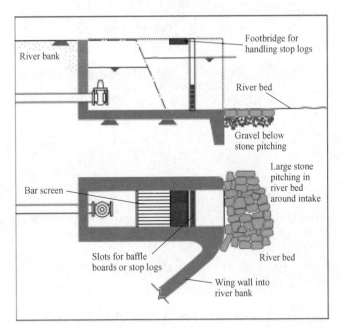

Fig. 23　Protected side intake

quired level, a sand trap to let the sand settle, and a spillway to release excess water.

(6) River-bottom intakes

River-bottom intakes for drinking-water systems are usually used in small rivers and streams where the sediment content and bed load transport are low (Fig. 24). The water is abstracted through a screen over a canal (usually made of concrete and built into the river bed). The bars of the screen are laid in the direction of the current and sloping downwards, so that coarse material cannot enter. From the canal, water enters a sand trap and then may pass a valve and flow by gravity, or be pumped into the rest of the system.

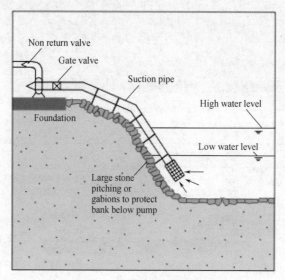

Fig. 24　River-bottom intake

(7) Floating intakes

Floating intakes (Fig. 25) for drinking-water systems allow water to be abstracted from near the surface of a river or lake, thus avoiding the heavier silt loads that are transported closer to the bottom during floods. The inlet pipe of a suction pump is connected just under the water level to a floating pontoon that is moored to the bank or bottom of the lake or river. The pump itself can be located either on the bank or on the pontoon. The advantages of placing the pump on the pontoon are that the suction pipe can be quite short and the suction head will be constant (less risk of cavitation).

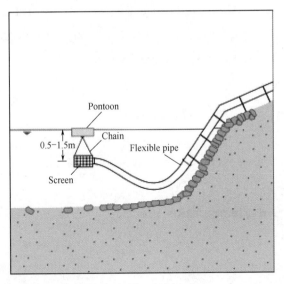

Fig. 25　Floating intake

(8) Sump intake

In a sump intake (Fig. 26), water from a river or lake flows through an underwater pipe to a well or sump from where it is lifted, usually into the initial purification stages of a drinking-water system. The inflow opening of the underwater pipe is located below the low-water lev-

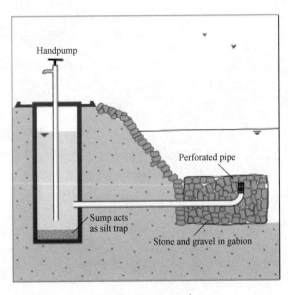

Fig. 26　Sump intake

el and is screened. A well provides a place for sedimentation to settle and protects the pump against damage by floating objects. To facilitate cleaning, two sump intakes are sometimes built for one pump.

3.5.2　Key points for operation

a. Types of water intake facilities;

b. Design guidelines of water intake systems;

c. Proposes standards for water intake facilities structures at existing facilities.

3.5.3　Outside operation practice

a. Operation and management of water intake facilities;

b. Operation and management technical requirements;

c. Actors and their roles.

Chapter 4 Conventional drinking water treatment processes

Treatment for drinking water production involves the removal of contaminants from raw water to produce water that is pure enough for human consumption without any short term or long term risk of any adverse health effect. Substances that are removed during the process of drinking water treatment include suspended solids, bacteria, algae, viruses, fungi and minerals such as iron and manganese.

The processes involved in removing the contaminants include physical processes such as settling and filtration, chemical processes such as disinfection and coagulation, and biological processes such as slow sand filtration.

Measures taken to ensure water quality not only relate to the treatment of the water, but to its conveyance and distribution after treatment. It is therefore common practice to keep residual disinfectants in the treated water to kill bacteriological contamination during distribution.

World Health Organization (WHO) guidelines are a general set of standards intended to apply where better local standards are not implemented. More rigorous standards apply across Europe, the USA and in most other developed countries, followed throughout the world for drinking water quality requirements.

4.1 Conventional rapid sand filtration system

For the provision of safe drinking water, rapid sand filtration system require adequate pre-treatment (usually coagulation-flocculation), sedimentation and post-treatment (usually disinfection with chlorine). This type of filtration results in flexible and reliable performance, especially when treating variable or very turbid source water. Both construction and operation are cost-intensive. It is a relatively sophisticated process usually requiring power-operated pumps, regular backwashing or cleaning, and flow control of the filter outlet. Rapid sand filtration is common in developed countries for the treatment of large quantities of water where land is a strongly limiting factor, and where material, skilled labour, and continuous energy supply are available (Fig. 27).

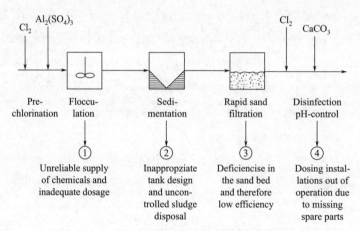

Fig. 27 Conventional rapid sand filtration system

4.2 Coagulation and flocculation

4.2.1 Fundamentals

Coagulation and flocculation are often the first steps in drinking water treatment (Fig. 28). Chemicals with a positive charge are added to the water. The positive charge of these chemicals neutralizes the negative charge of dirt and other dissolved particles in the water. When this occurs, the particles bind with the chemicals and form larger particles, called floc.

Coagulation is a water treatment process that causes very small sus-

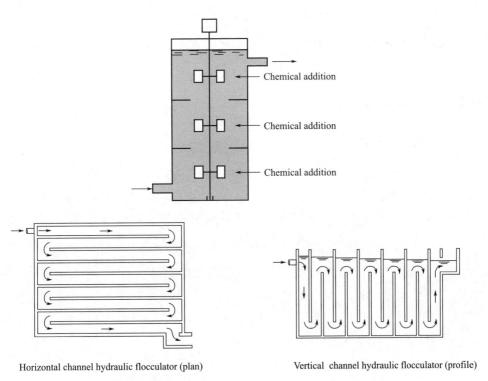

Fig. 28 Coagulation and flocculation

pended particles to attract to one another and form larger particles. Flocculation is a water treatment process following coagulation, which uses gentle stirring to bring the suspended particles together so they will form larger more settleable clumps called floc.

4.2.2 Key points for operation

a. Mechanical flash mixers: propeller type; turbine type; mechanical type; static type.

b. Rapid mix design considerations: coagulation occurs in two ways: by adsorption of soluble hydrolysis coagulant species on the colloid particles and destabilization by charge neutralization. These reactions occur in about 1 second; sweep floc coagulation where the coagulant exceeds its solubility limit and precipitates and traps the colloid particles. Sweep floc coagulation occurs in the range of 1-7 seconds.

4.2.3 Outside operation practice

a. Rapid mix tank design;

b. Flocculation practice in water treatment.

4.3 Sedimentation

4.3.1 Fundamentals

Sedimentation is a treatment process in which the velocity of the water is lowered below the suspension velocity and the suspended particles settle out of the water due to gravity. The process is known as settling or clarification. Most water treatment plants include sedimentation in their treatment processes. However, sedimentation may not be necessary in low turbidity water of less than 10 NTU. In this case, coagulation and flocculation are used to produce pinpoint (very small) floc which is removed from the water in the filters.

4.3.2 Key points for operation

(1) Location in the treatment process

The most common form of sedimentation follows coagulation and flocculation and precedes filtration. Sedimentation at this stage in the treatment process should remove 90% of the suspended particles from the water, including bacteria. The purpose of sedimentation here is to decrease the concentration of suspended particles in the water, reduce the load on the filters. Sedimentation can also occur as part of the pretreatment process, where it is known as pre-sedimentation. Pre-sedimentation can also be called plain sedimentation because the process depends merely on gravity and includes no coagulation and flocculation.

Without coagulation/flocculation, plain sedimentation can remove only coarse suspended matter (such as grit) which will settle rapidly out of the water without the addition of chemicals. This type of sedimentation typically takes place in a reservoir, grit basin, debris dam, or sand trap at the beginning of the treatment process.

(2) Types of basins

Rectangular basins are the simplest design, allowing water to flow horizontally through a long tank. This type of basin is usually found in large-scale water treatment plants. Rectangular basins have a variety of advantages-predictability, cost-effectiveness, and low maintenance. In addition, rectangular basins are the least likely to short-circuit, especially if the length is at least twice the width. A disadvantage of rectangular basins is the large amount of land area required.

Double-deck rectangular basins are essentially two rectangular sedimentation basins stacked one atop the other. This type of basin conserves land area, but has higher operation and maintenance costs than a one-level rectangular basin.

Square or circular sedimentation basins with horizontal flow are often known as clarifiers. This type of basin is likely to have short-circuiting problems.

(3) Zones

All sedimentation basins have four zones: the inlet zone, the settling zone, the sludge zone, and the outlet zone. Each zone should provide a smooth transition between the zone before and the zone after. In addition, each zone has its own unique purpose. Zones can be seen most easily in a rectangular sedimentation basin, such as the one shown below (Fig. 29):

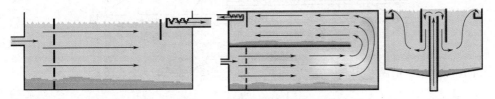

Fig. 29　Zone in a rectangular sedimentation basin

In a clarifier, water typically enters the basin from the center rather than from one end and flows out to outlets located around the edges of the basin. But the four zones can still be found within the clarifier (Fig. 30):

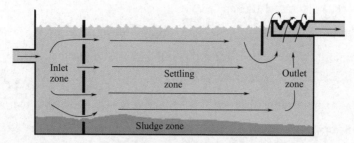

Fig. 30 Zone in a clarifier

4.3.3 Outside operation practice

a. Different sedimentation clarifier designs.

b. Assessment of main process characteristics:

Type 1-Dilute, non-flocculent, free-settling (every particle settles independently);

Type 2-Dilute, flocculent (particles can flocculate as they settle);

Type 3-Concentrated suspensions, zone settling, hindered settling (sludge thickening);

Type 4-Concentrated suspensions, compression (sludge thickening).

c. Settling of discrete particles.

d. Settlement of flocculent particles (Fig. 31).

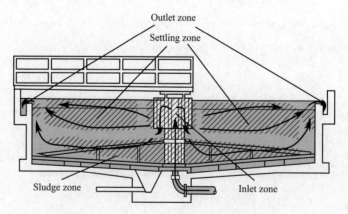

Fig. 31 Settlement of flocculent particles

e. Zone-settling behaviour.

f. Compression settling. The settling particles can contact each other and arise when approaching the floor of the sedimentation tanks at very high particle concentration. So that further settling will only occur in adjust matrix as the sedimentation rate decreasing. This is can be illustrated by the lower region of the zone-settling diagram below (Fig. 32). In compression zone, the settled solids are compressed by gravity (the weight of solids), as the settled solids are compressed under the weight of overlying solids, and water is squeezed out while the space gets smaller.

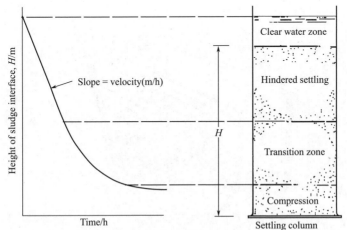

Fig. 32 Typical batch-settling column test on a suspension exhibiting zone-settling characteristics

4.4 Rapid sand filtration

4.4.1 Fundamentals

Types of Filters: slow sand filters; rapid sand filters.

Slow sand filters: the slow sand filter is the oldest type of large-scale filter (Fig. 33). The sand removes particles from the water through adsorption, straining, biological actions, with the formation of bio-film is the core.

Rapid sand filters: much greater filtration rate than SSF and the ability to clean automatically using backwashing. Rapid sand filters do not use biological filtration and depend primarily on adsorption and some straining (Fig. 34).

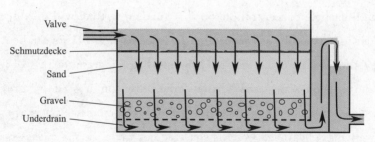

Fig. 33 Slow sand filters

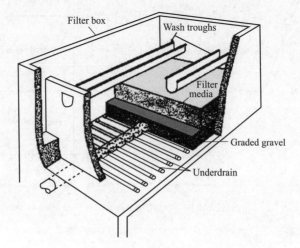

Fig. 34 Rapid sand filters

4.4.2 Key points for operation

(1) Factors influencing efficiency

To a large extent, the efficiency is determined by the characteristics of the water being treated and by the efficiency of previous stages in the treatment process. The chemical characteristics of the water being treated can influence both the preceding coagulation/flocculation and the filtration process. In addition, the characteristics of the particles in the water are especially important to the filtration process. Size, shape, and chemical characteristics of the particles will all influence filtration. For example, floc which is too large will clog the filter rapidly, requiring frequent backwashing, or can break up and pass through the filter, decreasing water quality. The type of filter used and the operation of the filter will influence filter efficiency.

(2) Mechanisms of removal by filtration

How are particles removed from water using filtration? Four mechanisms have been found to be part of the filtration process: straining, adsorption, biological action, and absorption.

4.4.3 Outside operation practice

a. Filtration process modeling;

b. Filter design;

c. When backwashing should be conducted;

d. The process of backwashing.

4.5 Disinfection

4.5.1 Fundamentals

Water disinfection means the removal, deactivation or killing of pathogenic microorganisms. Microorganisms are destroyed or deactivated, resulting in termination of growth and reproduction. Disinfection of drinking-water is essential if we are to protect the public from outbreaks of waterborne infectious and parasitic diseases. The main disinfectants evaluated in the Guidelines are free chlorine, chloramines, chlorine dioxide and ozone.

(1) Disinfection mechanism

Disinfection commonly takes place because of cell wall corrosion in the cells of microorganisms, or changes in cell permeability, protoplasm or enzyme activity (because of a structural change in enzymes). These disturbances in cell activity cause microorganisms to no longer be able to multiply. This will cause the microorganisms to die out. Oxidizing disinfectants also demolish organic matter in the water, causing a lack of nutrients (Fig. 35).

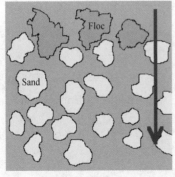
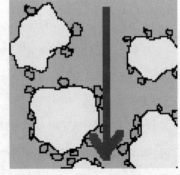

Straining: when $d_P > d_G$　　　　　　Adsorption: accumulation on granular surface

Fig. 35　Disinfection mechanism

(2) Media

Disinfection can be attained by means of physical or chemical disinfectants. The agents also remove organic contaminants from water, which serve as nutrients or shelters for microorganisms. Disinfectants should not only kill microorganisms. Disinfectants must also have a residual effect, which means that they remain active in the water after disinfection. A disinfectant should prevent pathogenic microorganisms from growing in the plumbing after disinfection, causing the water to be re-contaminated.

a. For chemical disinfection of drinking water the following disinfectants can be used:

-Chlorine (Cl_2);

-Chlorine dioxide (ClO_2);

-Hypo chlorite (OCl^-);

-Ozone (O_3).

b. For physical disinfection of drinking water the following disinfectants can be used:

-Ultraviolet light (UV);

-Electronic radiation;

-Gamma rays;

-Sounds;

-Heat.

4.5.2 Key point for operation

Balancing chemical and microbial risks: quantitative assessments of risks associated with the microbial contamination of drinking water are scarce. Although there are gaps in our knowledge, we can not afford to postpone action until rigorous quantitative assessment of chemical versus microbial risks are available and every answer is known.

4.5.3 Outside operation practice

Chemical inactivation of microbiological contamination in natural or untreated water is usually one of the final steps to reduce pathogenic microorganisms in drinking water. Combinations of water purification steps (oxidation, coagulation, settling, disinfection, filtration) cause drinking water to be safe after production. As an extra measure, many countries apply a second disinfection step at the end of the water purification process, in order to protect the water from microbiological contamination in the water distribution system. Usually one uses a different kind of disinfectant from the one earlier in the process, during this disinfection process. The secondary disinfection makes sure that bacteria will not multiply in the water during distribution. Bacteria can remain in the water after the first disinfection step or can end up in the water during backflushing of contaminated water (which can contain groundwater bacteria as a result of cracks in the plumbing).

Chapter 5 Water distribution

Water distribution systems consist of an interconnected series of components (Fig. 36). They include pipes, storage facilities and components that convey drinking water. The purpose of distribution system is to deliver water to consumer with appropriate quality, quantity and pressure. Distribution system is used to describe collectively the facilities used to supply water from its source to the point of usage.

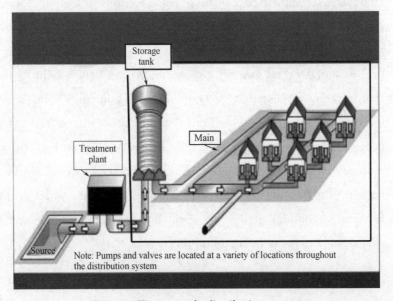

Fig. 36 Water supply distribution system

Public water systems depend on distribution systems to provide an uninterrupted supply of pressurized safe drinking water to all consumers. Distribution system mains carry water from either the treatment plant to the consumer or the source to the consumer when treatment is absent.

Distribution systems span almost one million miles in the United States. They represent the vast majority of physical infrastructure for water supplies. Distribution system wear and tear can pose intermittent or persistent health risks.

(1) Water quality and the distribution system

New pipes are added to distribution systems as development occurs. The additions result in a wide variation in pipe sizes, materials, methods of construction, age within individual distribution systems and across the nation.

As these systems age, deterioration can occur due to corrosion, materials erosion, and external pressures. Deteriorating water distribution systems can lead to breaches in pipes and storage facilities, intrusion due to water pressure fluctuation and main breaks.

(2) Protecting water quality in distribution systems

The following EPA drinking water regulations pertain to distribution systems:

a. Surface water treatment rules (disinfectant residual and sanitary survey requirements);

b. Stage 1 and Stage 2 disinfectants and disinfection byproducts rules (DBPR, monitoring for DBPs in the distribution system);

c. Ground water rule (sanitary surveys);

d. Total coliform rule (monitoring for bacterial contamination in distribution systems).

5.1 Distribution reservoirs

5.1.1 Fundamentals

Distribution reservoirs (Fig. 37), also called service reservoirs, are the storage reservoirs, which store the treated water for supplying water during emergencies (such as during fires, repairs, etc.) and also to help in absorbing the hourly fluctuations in the normal water demand. The service reservoir is provided to balance the (constant) supply rate from the water source/treatment plant with the fluctuating water demand in the distribution area. The storage volume should be large enough to accommodate the cumulative differences between water supply and demand. Without storage of water in the distribution area, the source of supply and the water treatment plant would have to be able to follow all fluctuations in the water demand of the community served. This is generally not economical, and sometimes not even technically feasible. Schematic of distribution reservoirs is shown in Fig. 38.

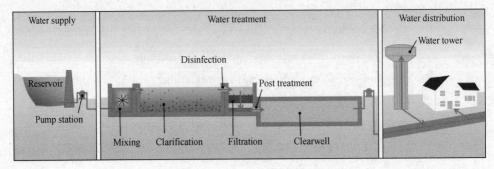

Fig. 37　Distribution reservoirs

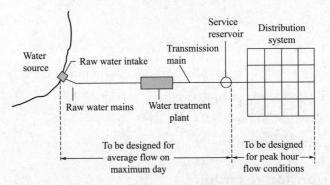

Fig. 38　Schematic of distribution reservoirs

5.1.2　Key point for operation

(1) Functions of distribution reservoirs

a. To absorb the hourly variations in demand;

b. To maintain constant pressure in the distribution mains;

c. Water stored can be supplied during emergencies.

(2) Location and height of distribution reservoirs

a. Should be located as close as possible to the center of demand;

b. Water level in the reservoir must be at a sufficient elevation to permit gravity flow at an adequate pressure.

(3) Types of reservoirs

a. Underground reservoirs;

b. Small ground level reservoirs;

c. Large ground level reservoirs;

d. Overhead tanks.

(4) Storage capacity of distribution reservoirs

The total storage capacity of a distribution reservoir is the summation of:

a. Balancing storage: the quantity of water required to be stored in the reservoir for equalizing or balancing fluctuating demand against constant supply is known as the balancing storage (or equalizing or operating storage). The balance storage can be worked out by mass curve method.

b. Breakdown storage: the breakdown storage or often called emergency storage is the storage preserved in order to tide over the emergencies posed by the failure of pumps, electricity, or any other mechanism driving the pumps. A value of about 25% of the total storage capacity of reservoirs, or 1.5-2 times of the average hourly supply, may be considered as enough provision for accounting this storage.

c. Fire storage: the third component of the total reservoir storages is the fire storage. This provision takes care of the requirements of water for extinguishing fires. A provision of 1-4 per person per day is sufficient to meet the requirement.

The total reservoir storages can finally be worked out by adding all the three storages.

5.1.3 Outside operation practice

a. Distribution system design;

b. Protecting water quality in distribution systems;

c. Distribution system problems and recommendations on reducing risk.

5.2 Distribution pipe system

5.2.1 Fundamentals

A water pipe is any pipe or tube designed to transport treated drinking water to consumers. In well planned and designed water distribution networks, water is generally treated before distribution and sometimes also chlorinated, in order to prevent recontamination on the way to the end user. The varieties include large diameter main pipes, which supply entire towns, smaller branch lines that supply a street or group of buildings, or small diameter pipes located within individual buildings. Materials commonly used to construct water pipes include cast iron, polyvinyl chloride (PVC), copper, steel or concrete.

5.2.2 Key point for operation

Types of pipes: pipes come in several types and sizes. They can be divided into three main categories: metallic pipes, cement pipes and plastic pipes. Metallic pipes include steel pipes, galvanized iron pipes and cast iron pipes. Cement pipes include concrete cement pipes and asbestos cement pipes. Plastic pipes include plasticized polyvinyl chloride (PVC) pipes.

a. Steel pipes are comparatively expensive, but they are the strongest and most durable of all water supply pipes.

b. Galvanized steel or iron pipes are the traditional piping materials in the plumbing industry for the conveyance of water and wastewater.

c. Cast iron pipes are quite stable and well suited for high water pressure. However, cast iron pipes are heavy, which makes them unsuitable for inaccessible places due to transportation problems. In addition, due to their weight they generally come in short lengths increasing costs for layout and jointing.

d. Concrete cement and asbestos cement pipes are expensive but non-corrosive by nature. Their advantage is that they are extremely strong and durable. However, being bulky and heavy, they are harder and more costly to handle, install and transport.

e. Plasticized Polyvinyl Chloride (PVC) pipes are non-corrosive, extremely light and thus easy to handle and transport. Still, they are strong and come in long lengths that lower installation/transportation costs. However, they are prone to physical damage if exposed overground and become brittle when exposed to ultraviolet light. In addition to the problems associated with the expansion and contraction of PVC, the material will soften and deform if exposed to temperatures over 65℃.

5.2.3 Outside operation practice

a. Weight of the pipe;

b. Ease of assembling;

c. Pipe strength;

d. Health aspects.

5.3 Pumping stations

5.3.1 Fundamentals

Pumping stations (Fig. 39) in a water distribution system are necessary where water is pumped directly into the system (e.g. from a lake) or where pressure has to be increased because there is an insufficient difference in water levels in gravity flow distribution systems. There are two general types of pumps: vertical turbine pumps and centrifugal pumps. Capital costs are high, but the most expensive part is the energy supply for pumps (mostly electrical). Therefore, it is very important that pumps have a high degree of efficiency and are maintained properly. To guarantee safe water quality, cross connection of drinking water and waste removal systems must be avoided.

Fig. 39　Pumping station

The basic design principles are as follows.

Main pumping stations, which supply water to the distribution system, are located near the water treatment facility or a potable water storage facility, and pump directly into the piping system. Pumps that pump directly into transmission lines and distribution systems are sometimes called high lift pumps.

Booster pumps are additional pumps used to increase pressure locally or temporarily. Booster pumps stations are usually remotely located from the main pump station, as in hilly topography where high-pressure zones are required, or to handle peak flows in a distribution system that can otherwise handle the normal flow requirements.

Where a pump station is added to an existing installation, previous planning and design, which are based upon a total system hydraulic analysis, should be consulted before the addition is designed.

The sizing of each component in the distribution system will depend upon the effective combination of the major system elements:

a. Supply source;

b. Storage (e. g. in reservoirs);

c. Distribution piping;

d. Pumping.

The location of the pump station and intake structure, and the anticipated heads and capacities are the major factors in the selection of pumps. The function of a pump station in the overall distribution system operation can also affect the determination of capacities.

5.3.2 Key points for operation

(1) Pump types

There are generally two types of pumps used for potable water pumping applications:

a. The vertical turbine pump (line shaft and submersible types);

b. The centrifugal horizontal or vertical split case pump designed for water-works service.

(2) Pump discharge capacity

a. If the pump is used directly to supply water without a reservoir, the capacity must be equal to the peak hour demand;

b. If the water distribution system has a reservoir, the pump capacity must be equal to the maximum daily demand.

(3) Pump selection

a. If the pumping water level is less than 6 meters, use a centrifugal pump (maximum suction lift=6 meters);

b. If the pumping water level is from 6 to 20 meters, use jet pumps or a submersible;

c. If the pumping water level is greater than 20 meters, use a submersible or a vertical line shaft turbine pump.

(4) Power supply for pumps

Electric, gasoline or diesel engines are commonly used as power sources for pumps. The electric motor is, however, the most favoured power source because of its reliability, relatively low power cost, and environmental considerations like cleanliness, relatively low noise, and low pollutant emissions. An electrical pump may also be driven with solar power. Heat sensors installed in the windings during manufacturing should protect electric motor. These sensors shut the motor off in case of low voltage or change in phase before damage can be done.

5.3.3 Outside operation practice

a. Cost consideration.

b. Health aspects: a cross-connection of drinking water distribution and waste removal system must be avoided. This risk is usually greater in public, industrial and commercial premises, where dual water systems, circulating pumps, toxic wastes and other factors have to be managed. There are also risks in multi-storey buildings where booster pumps are employed to increase the mains pressure, and in special systems used in hospitals and in dental and veterinary surgeries. However, even standard single-family domestic buildings present health risks to both occupants and neighbours if faulty plumbing is installed or if plumbing is not maintained.

c. Operation and maintenance.

Chapter 6 Miscellaneous water treatment techniques

Fundamentals of water treatment:

a. Slow sand filtration;

b. Sedimentation, coagulation, flocculation, settling, fast sand filtration;

c. Softening approaches: lime softening, membrane softening;

d. Filtration: membrane filtration (ultrafiltration, microfiltration);

e. Disinfection: chlorine, UV, ozone, chlorine dioxide;

f. On-site generation.

6.1 Advanced oxidation (ozonation, UV radiation and hydroxyl oxidation)

6.1.1 Fundamentals

Advanced oxidation is chemical oxidation with hydroxyl radicals, which are very reactive, and short-lived oxidants. Advanced oxidation processes (AOPs) refer to a set of oxidative water treatments that can be used to treat toxic effluents at industrial level, hospitals and wastewater treatment plants (Fig. 40).

AOPs in general are cheap to install but involve high operating costs due to the input of chemicals and energy required. To limit the costs, AOPs are often used as pre-treatment combined with biologic treatment. Advanced oxidation was recently also used as a polishing step to remove micro-pollutants from the effluents of municipal wastewater treatment plants and for the disinfection of water.

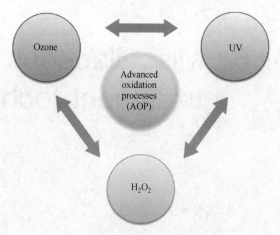

Fig. 40　Advanced oxidation processes

6.1.2　AOPs Mechanism

Advanced oxidation involves several steps schematized in Fig. 41 and explained as follows:

a. Formation of strong oxidants (e.g. hydroxyl radicals);

b. Reaction of these oxidants with organic compounds in the water producing biodegradable intermediates;

c. Reaction of biodegradable intermediates with oxidants referred to as mineralization (i.e. production of water, carbon dioxide and inorganic salts).

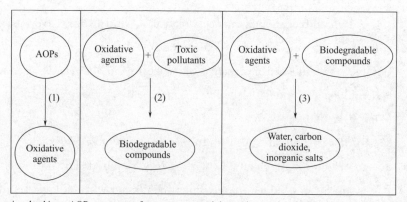

Main steps involved in on AOPs treatment of wastewater containing toxic organic compounds, Source:MAZILLE (2011)

Fig. 41　AOPs mechanism

Types of advanced oxidation process are shown in Fig. 42.

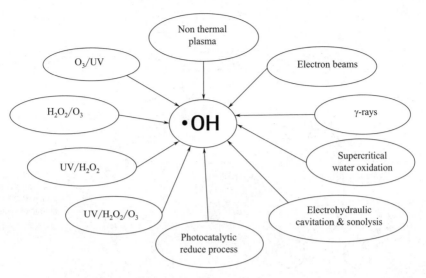

Fig. 42 Types of advanced oxidation process

6.1.3 Ozonation

Ozone is a powerful oxidant (far more so than dioxygen) and has many industrial and consumer applications related to oxidation. This same high oxidizing potential, however, causes ozone to damage mucous and respiratory tissues in animals, and also tissues in plants, above concentrations of about 100ppb. This makes ozone a potent respiratory hazard and pollutant near ground level. However, the ozone layer (a portion of the stratosphere with a higher concentration of ozone, from 2 to 8 ppm) is beneficial, preventing damaging ultraviolet light from reaching the Earth's surface, to the benefit of both plants and animals. Fig. 43 shows the oxygen-ozone-oxygen cycle.

(1) Key points for operation

a. Ozone as disinfectant. Ozone is a well-known powerful oxidizer which could kill microorganisms effectively. Ozone can only be used as a primary disinfectant and it cannot maintain a residual in the distribution system, and its residual decays too rapidly. Thus, ozone disinfection should be coupled with a secondary disinfectant, such as chlorine, chloramine, or chlorine dioxide for a complete disinfection system.

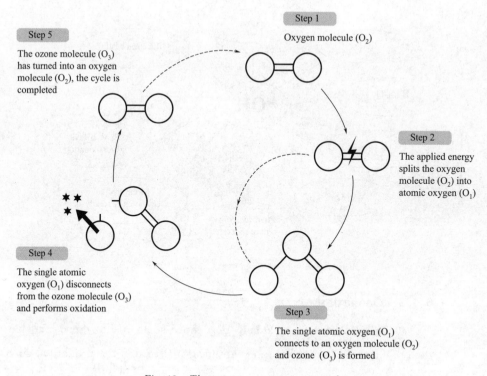

Fig. 43 The oxygen-ozone-oxygen cycle

b. Inactivation mechanisms of bacteria (Fig. 44).

c. Inactivation mechanisms of viruses (Fig. 44).

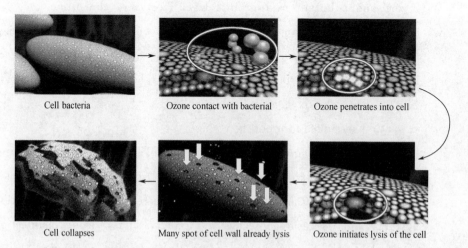

Fig. 44 Inactivation mechanisms of bacteria

(2) Outside operation practice

a. Effectiveness of ozone disinfection;

b. Ozonation disinfection byproducts;

c. Ozone byproducts control.

6.1.4 UV radiation

Ultraviolet (UV) is an electromagnetic radiation with a wavelength from 10 nm (30 PHz) to 400 nm (750 THz), shorter than that of visible light but longer than X-rays. UV radiation constitutes about 10% of the total light output of the Sun, and is thus present in sunlight. It is also produced by electric arcs and specialized lights, such as mercury-vapor lamps, tanning lamps, and black lights. Although it is not considered an ionizing radiation because its photons lack the energy to ionize atoms, long-wavelength ultraviolet radiation can cause chemical reactions and cause many substances to glow or fluoresce. Consequently, the biological effects of UV are greater than simple heating effects, and many practical applications of UV radiation derive from theirs interactions with organic molecules (Fig. 45).

Fig. 45 UV radiation

Simply put, ultraviolet radiation (also known as UV radiation or ultraviolet rays) is a form of energy traveling through space. Some of the most frequently recognized types of energy are heat and light. These, along with others, can be classified as a phenomenon known as electromagnetic radiation. Other types of electromagnetic radiation are gamma rays, X-rays, visible light, infrared rays, and radio waves. The progression of electromagnetic radiation through space can be visualized in different ways. Some experiments suggest that these rays travel in the form of waves. A physicist can actually measure the length of those waves (simply called their wavelength). It turns out that a smaller wavelength means more energy. At other times, it is more plausible to describe electromagnetic radiation as being contained and traveling in little packets, called photons.

The distinguishing factor among the different types of electromagnetic radiation is their energy content. Ultraviolet radiation is more energetic than visible radiation and therefore has a shorter wavelength. To be more specific: ultraviolet rays have a wavelength between approximately 100nm and 400nm whereas visible radiation includes wavelengths between 400nm and 780nm.

(1) Key points for operation

a. How is radiation classified on the electromagnetic spectrum (Fig. 46).

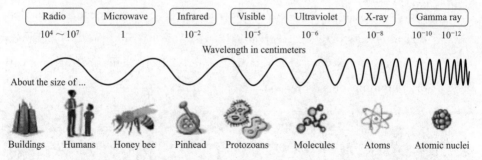

Fig. 46 Radiation classification on the electromagnetic spectrum

Electromagnetic radiation is all around us, though we can only see some of it. All EM radiation (also called EM energy) is made up of minute packets of energy or "particles", called photons, which travel in a wave-like pattern and move at the speed of light. The EM spectrum is divided into categories defined by a range of numbers. These ranges describe the activity level, or how energetic the photons are, and the size of the wavelength in each category.

For example, at the bottom of the spectrum radio waves have photons with low energies, so their wavelengths are long with peaks that are far apart. The photons of microwaves have higher energies, followed by infrared waves, UV rays, and X-rays. At the top of the spectrum, gamma rays have photons with very high energies and short wavelengths with peaks that are close together.

(2) Outside operation practice

a. What are the different types of UV radiation?

b. Are there health benefits of exposure to UV radiation?

6.1.5 Hydroxyl radicals

The hydroxyl radical, ·OH, is the neutral form of the hydroxide ion (OH^-) (Fig. 47). Hydroxyl radicals are highly reactive (easily becoming hydroxyl groups) and consequently short-lived; however, they form an important part of radical chemistry. Most notably hydroxyl radicals are produced from the decomposition of hydroperoxides (ROOH) or, in atmospheric chemistry, by the reaction of excited atomic oxygen with water. It is also an important radical formed in radiation chemistry, since it leads to the formation of hydrogen peroxide and oxygen, which can enhance corrosion and SCC in coolant systems subjected to radioactive environments. Hydroxyl radicals are also produced during UV-light dissociation of H_2O_2 and likely in Fenton chemistry, where trace amounts of reduced transition metals catalyze peroxide-mediated oxidations of organic compounds.

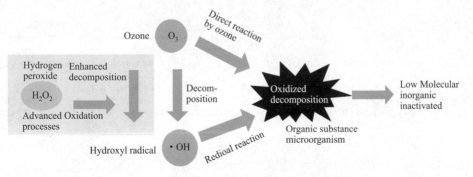

Fig. 47　Hydroxyl radicals

(1) Key points for operation

a. Production and contribution of hydroxyl radicals;

b. Atmospheric importance;

c. Biological significance.

(2) Outside operation practice

a. Hydroxyl radicals detections;

b. Hydroxyl radicals formation;

c. Hydroxyl radicals and its scavengers in health and disease.

6.2 Activated carbon adsorption

6.2.1 Fundamentals

Adsorption is a mass transfer operation in which substances present in a liquid phase are adsorbed or accumulated on a solid phase and thus removed from the liquid (Fig. 48).

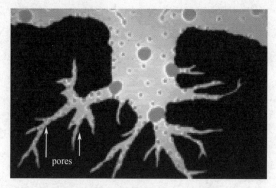

Fig. 48 Schematic of adsorption

Adsorption process is used in drinking water treatment for removal of taste-causing and odor-causing compounds, synthetic organic chemicals (SOCs), color-forming organics and disinfection byproducts (DBP) and its precursors. Inorganic constituents, including some that represent a health hazard, such as arsenic and some heavy metals, can also be removed by adsorption.

6.2.2 Key points for operation

a. Types of adsorbents.

Several different materials can be used as adsorbents in water treatment. The most widespread of these materials is activated carbon which is formed when carbon from wood, coal, peat, or nut shells is carbonized and activated by exposing to heat in the absence of oxygen (Fig. 49).

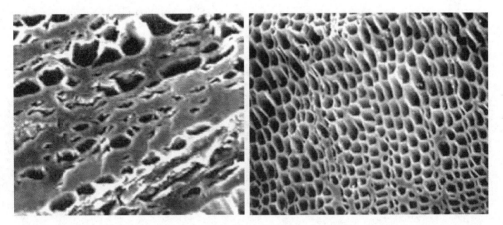

Fig. 49   Activated carbon pores

b. Atmospheric importance.

c. Biological significance.

The other two types of adsorbents are activated alumina and synthetic resins, both of which are typically used as filter media. Activated alumina is used to remove excess fluoride from water as well as to remove arsenic and selenium. Synthetic resins can remove trihalomethanes from water. However, synthetic resins are very costly and theirs use is still in the developmental stages.

Fig. 50 shows the adsorption process.

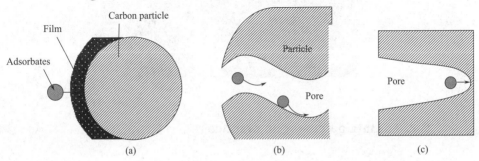

Fig. 50   Adsorption process

Chapter 6   Miscellaneous water treatment techniques   77

d. Types of activated carbon and application.

-Powdered activated carbon (Fig. 51): powdered activated carbon, PAC, a form of activated carbon with a very small particle size. Treatment involves adding PAC to water, allowing the PAC to interact with contaminants in the water, then removing the PAC by sedimentation or filtration.

Fig. 51　Powdered activated carbon

-Granular activated carbon (Fig. 52): also known as GAC, has a larger particle size than PAC with an associated greater surface area. Like PAC, GAC can remove trihalomethane precursors as well as taste-causing and odor-causing compounds. GAC is used as a filter medium, either as a layer in a rapid-sand filter or in a separate filter known as an adsorber or a contactor. When adsorbers are used, the contactor is placed downstream of the filter so that turbidity won't clog the adsorber.

Fig. 52　Granular activated carbon

6.2.3　Outside operation practice

a. Choosing a type of activated carbon.

GAC and PAC each have advantages and disadvantages. In general, PAC is used more often due to the low initial cost and to the flexibility of dosage which allows the PAC concentration to be adjusted to deal with changing contaminant levels. However, PAC has a high operating cost if used continuously, cannot be regenerated, produces large quantities of sludge, and can break through filters to cause dirty water complaints by the customers. In addition, the dust resulting from the small particles of PAC make handling difficult. GAC becomes a more economical choice in larger systems or where taste and odor must be controlled continuously. Disadvantages of GAC include a high initial cost to construct the adsorber or contactor, and the tendency of GAC filters to grow bacteria.

b. Efficiency assessment (in case powdered activated carbon is used; in case granular activated carbon is used).

c. Application.

Fig. 53 shows an application example in the USA.

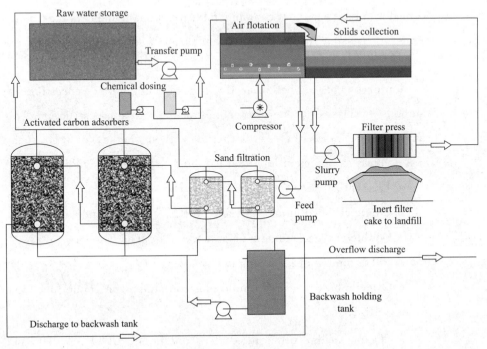

Fig. 53　Application example (USA)

6.3 Ion exchange

6.3.1 Fundamentals

Ion exchange is an exchange of ions between two electrolytes or between an electrolyte solution and a complex. In most cases the term is used to denote the processes of purification, separation, and decontamination of aqueous and other ion-containing solutions with solid polymeric or mineralic "ion exchangers".

Typical ion exchangers are ion exchange resins (functionalized porous or gel polymer), zeolites, montmorillonite, clay, and soil humus. Ion exchangers are either cation exchangers that exchange positively charged ions (cations) or anion exchangers that exchange negatively charged ions (anions). There are also amphoteric exchangers that are able to exchange both cations and anions simultaneously. However, the simultaneous exchange of cations and anions can be more efficiently performed in mixed beds that contain a mixture of anion and cation exchange resins, or passing the treated solution through several different ion exchange materials.

Ion exchanges can be unselective or have binding preferences for certain ions or classes of ions, depending on their chemical structure. This can be dependent on the size of the ions, their charge, or their structure. Typical examples of ions that can bind to ion exchangers are:

a. H^+ (proton) and OH^- (hydroxide);

b. Single-charged monatomic ions like Na^+, K^+, and Cl^-;

c. Double-charged monatomic ions like Ca^{2+} and Mg^{2+};

d. Polyatomic inorganic ions like SO_4^{2-} and PO_4^{3-};

e. Organic bases, usually molecules containing the amine functional group—NR_2H^+;

f. Organic acids, often molecules containing —COO^- (carboxylic acid) functional groups;

g. Biomolecules that can be ionized: amino acids, peptides, proteins, etc.

Along with absorption and adsorption, ion exchange is a form of sorption.

Ion exchange is a reversible process and the ion exchanger can be regenerated or loaded with desirable ions by washing with an excess of these ions.

6.3.2 Key points for operation

(1) Ion exchange treatment of drinking water

Ion exchange is a water treatment method where one or more undesirable contaminants are removed from water by exchange with another non-objectionable, or less objectionable substance. Both the contaminant and the exchanged substance must be dissolved and have the same type (+, −) of electrical charge. One example of ion exchange is the process called "water softening".

(2) The Exchange of ions

The electrical charge on an ion can be either positive (+) or negative (−). Valence is the term that describes the category of the electrical charge on a dissolved ion such as positive 2 or positive 3. If the contaminant has a positive charge, it would be called a cation, and would be removed by use of an IE media called a cation exchange resin. If the contaminant has a negative charge, it would be called an anion, and the appropriate treatment media would be called an anion exchange resin.

(3) The water softening process

A water softener at a private home typically has two or three tanks. The smaller tank contains the sodium or potassium salt used to regenerate the resin media while the taller tank (s) contains the purifying media called a "cation" exchange resin (Fig. 54). During normal operations, raw water passes through the ion exchange resin media in the tall tank. The calcium (Ca^{2+}), magnesium (Mg^{2+}), iron (Fe^{2+}), or manganese (Mn^{2+}) ions in the water are typically "exchanged" for sodium (Na^+) or potassium (K^+) ions, which have been temporarily stored in the pores of the resin during the previous regeneration cycle. In fact, any contaminant ion of valence positive 2 or greater will be removed in a water softener.

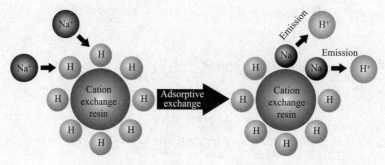

Fig. 54 Ion exchange resins

As the softener removes hardness minerals from the water, sodium or potassium will be given back proportionally. Shown below (Table 9) is the concentration of either sodium or potassium that would be added to the existing raw water if 10 mg/L of hardness is removed.

Table 9 Concentration of sodium and potassium added to the existing raw water if 10 mg/L of hardness is removed

Hardness Removed	Na^+ or K^+ Added	Concentration
10mg/L as $CaCO_3$	Sodium(Na^+) added	4.6 mg/L
10mg/L as $CaCO_3$	Potassium(K^+) added	7.6 mg/L

If we think of atoms and balanced electrical charges: then if we have 10 atoms (ions) of hardness [calcium (Ca^{2+}), magnesium (Mg^{2+})], we will be adding 20 atoms (ions) of sodium (Na^+) or potassium (K^+).

6.3.3 Outside operation practice

a. Choosing a type of activated carbon;

b. Categorizing hardness;

c. Expressing the amount of hardness in water;

    d. Initiation of the regeneration cycle;

    e. Strength of brine used to regenerate;

    f. Capture last part brine backwash;

    g. Disposal of wastes to leach field.

6.4 Membrane filtration

6.4.1 Fundamentals

Widely recognized as the technology of choice for superior water and wastewater treatment, membranes provide a physical barrier that effectively removes solids, viruses, bacteria and other unwanted molecules. Different types of membranes are used for softening, disinfection, organic removal, and desalination of water and wastewater and can be installed in compact, automated, modular units. Membrane filtration units can also be installed in relatively small facilities that blend into the surrounding area and can be fully automated to significantly reduce the required amount of operator attention. Fig. 55 shows the mechanism of membrane filtration.

Recent advances in technology have significantly reduced the cost of membrane-based systems. Installation costs are lower because membrane systems don't require large buildings or as much land as conventional systems. Operating costs are reduced since today's membranes produce more water and remove more impurities while using less energy.

In the United States, regulations such as the *Safe Drinking Water Act* and the *Long Term 2 Enhanced Surface Water Treatment Rule* have had a significant impact on municipal water treatment. This, in addition to increasingly stringent wastewater discharge regulations, has promoted dramatic growth in the implementation of membrane technology.

6.4.2 Key points for operation

Membranes provide physical barriers that permit the passage of materials only up to a certain size, shape or character. There are four cross-

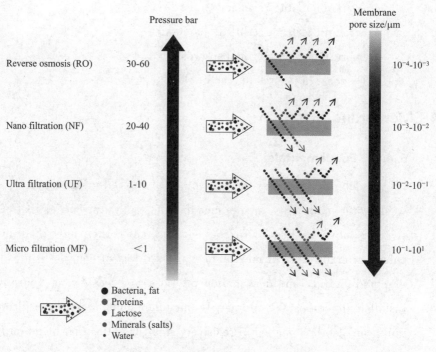

Fig. 55 Mechanism of membrane filtration

flow, pressure-driven membrane separation processes currently employed for liquid/liquid and liquid/solid separation: ultrafiltration (UF), reverse osmosis (RO), nanofiltration (NF), and microfiltration (MF). Membranes are manufactured in a variety of configurations including hollow fiber, spiral, and tubular shapes. Each configuration offers varying degrees of separation.

a. Ultrafiltration: ultrafiltration (UF) is a pressure-driven process that removes emulsified oils, metal hydroxides, colloids, emulsions, dispersed materials, suspended solids, and other large molecular weight materials from water. UF membranes are characterized by their molecular weight cut-off. The major opportunities for UF involve clarification of solutions containing suspended solids, removal of viruses and bacteria or high concentrations of macromolecules. These include oil/water separation, fruit juice clarification, milk and whey production and processing, automotive electrocoat paint filtration, purification of pharmaceuticals, poly-vinyl alcohol and indigo recovery, potable water production, and secondary or tertiary wastewater reuse.

b. Reverse osmosis: the membrane with the smallest pores is reverse osmosis (RO), which involves reversal of the osmotic process of a solution in order to drive water away from dissolved molecules. RO depends on ionic diffusion to effect the separation. A common application of RO is seawater and brackish water desalination. RO is also used in many industrial processes including cheese whey concentration, fruit juice concentration, ice-making, and car wash water reclamation, and wastewater volume reduction. In each of these examples, the goal is either to produce a pure filtrate (typically water), reduce the volume of the wastewater for disposal or retain the components of the feed stream as the product.

c. Nanofiltration: nanofiltration (NF) functions similarly to RO, but is generally targeted to remove only divalent and larger ions. Monovalent ions such as sodium and chloride will pass through an NF membrane, thus many of the uses of NF involve de-salting of the process stream. An example is the production of lactose from cheese whey; the NF process is designed to concentrate the lactose molecules while passing salts-a procedure that purifies-concentrates-the lactose stream. In water treatment, NF membranes are used for hardness removal (in place of water softeners), pesticide elimination and color reduction. Nanofiltration can also be used to reclaim spent NaOH solutions. In this case, the permeate (filtrate) stream is purified NaOH, allowing reuse many times over.

d. Microfiltration: microfiltration (MF) has significant applications in simple dead-end filtration for water filtration, sterile bottling of fruit juices and wine, and aseptic uses in the pharmaceutical industry. A large portion of the MF market has been captured by crossflow. The most common of these are clarification of whole cell broths and purification processes in which macromolecules must be separated from other large molecules, proteins and/or cell debris. Clarification of dextrose and highly-colored fruit juices employ MF extensively as well. There are also large markets for MF crossflow filtration in wine production, milk/whey de-fatting and brewing. MF systems operate at relatively low pressures and are configured based upon the application.

6.4.3 Outside operation practice

There are a variety of membrane technologies and configurations available in the marketplace, each of which provides certain advantages for specific process needs. The key to a successful membrane filtration system is to carefully evaluate the physical characteristics of the process fluid.

A membrane bioreactor (MBR) is a biological process that combines secondary and tertiary treatment using a membrane filtration process. MBR systems are growing in popularity for virtually all wastewater treatment applications because they offer many advantages over conventional wastewater treatment plants such as consistently high quality effluent with low turbidity, low bacterial counts, low TSS and NTU, while using fewer chemicals than conventional wastewater treatment plants. The filtrate quality, in many instances, is suitable for feeding directly into an RO process. An additional advantage of an MBR system is its compact footprint.

 a. Key points to consider when evaluating an MBR system;
 b. Membrane transport mechanism;
 c. Types of membrane processes;
 d. Dead end filtration: theories & equations.

6.5 Reverse osmosis

6.5.1 Fundamentals

Reverse osmosis or RO is a filtration method that is used to remove ions and molecules from a solution by applying pressure to the solution on one side of a semipermeable or selective membrane (Fig. 56). Large molecules (solute) can't cross the membrane, so they remain on one side. Water (solvent) can cross the membrane. The result is that solute molecules become more concentrated on one side of the membrane, while the opposite side becomes more dilute.

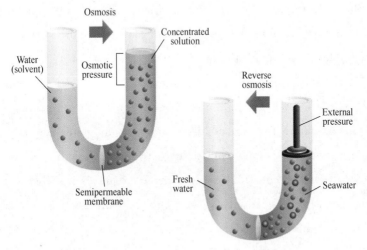

Fig. 56 Osmosis and reverse osmosis

6.5.2 Key points for operation

(1) How reverse osmosis works

In order to understand reverse osmosis, it helps to first understand how mass is transported via diffusion and regular osmosis. Diffusion is the movement of molecules from a region of higher concentration to a region of lower concentration. Osmosis is a special case of diffusion in which the molecules are water and the concentration gradient occurs across a semipermeable membrane. The semipermeable membrane allows the passage of water, but not ions (e.g. Na^+, Ca^{2+}, Cl^-) or larger molecules (e.g. glucose, urea, bacteria). Diffusion and osmosis are thermodynamically favorable and will continue until equilibrium is reached. Osmosis can be slowed, stopped, or even reversed if sufficient pressure is applied to the membrane from the "concentrated" side of the membrane.

Reverse osmosis occurs when the water is moved across the membrane against the concentration gradient, from lower concentration to higher concentration. To illustrate, imagine a semipermeable membrane with fresh water on one side and a concentrated aqueous solution on the other side. If normal osmosis takes place, the fresh water will cross the membrane to dilute the concentrated solution. In reverse osmosis, pressure is exerted on the side with the concentrated solution to force the water molecules through the membrane to the fresh water side.

There are different pore sizes of membranes used for reverse osmosis. While a small pore size does a better job of filtration, it takes longer to move water. It's sort of like trying to pour water through a strainer (large holes or pores) compared to trying to pour it through a paper towel (smaller holes). However, reverse osmosis is different from simple membrane filtration because it involves diffusion and is affected by flow rate and pressure.

(2) Uses of reverse osmosis

Reverse osmosis is often used in commercial and residential water filtration. It is also one of the methods used to desalinate seawater. Reverse osmosis not only reduces salt, but can also filter out metals, organic contaminants, and pathogens. Sometimes reverse osmosis is used to purify liquids in which water is an undesirable impurity. For example, reverse osmosis can be used to purify ethanol or grain alcohol to increase its proof.

6.5.3 Outside operation practice

a. Principles of operation;

b. Types of reverse osmosis membranes.

6.6 Biological filtration

6.6.1 Fundamentals

The successful aquarium is virtually a living ecosystem. A part of the ecosystem involves microorganisms and chemical reactions. A sound biological filter is considered by many, the most important filtration system of the aquarium ecosystem.

A biological filter is one that involves the propagation and retention of billions of aerobic and anaerobic bacteria. Most aquarists pay close attention to the aerobic, or nitrifying bacteria (*Nitrosomonas*, *Nitrobacter*). These bacteria grow in the presence of oxygenated water with a food source such as ammonia (NH_3), or nitrite (NO_2) present.

Initially, the sound of growing bacteria, and creating a biologically active ecosystem sounds like a complicated process, but it actually is fairly simple with the right conditions, equipment, and animals.

Most commonly, the bacteria are added by simply adding fish. They hitch a ride on the very bodies, mouths, and gills of fish and other living organisms. They will drop off the bodies of the fish, and spread throughout the aquarium. Anywhere there is oxygen-rich water, these bacteria will grow. The only problem with this process, is that fish also produce ammonia. In fact, more than the bacteria to compensate for its own waste. So fish need to be added gradually, because these bacteria can take several months to fully establish.

These nitrifying bacteria consume ammonia (NH_3) and nitrite (NO_2) to break these common toxic aquarium chemicals into nitrate (NO_3), the final bi-product. NO_3 is then removed either by further chemical processes, or more commonly, by changing out a percentage of the total aquarium water on a regular basis, and replacing with water that is free of heavy metals, and is the appropriate temperature, pH, and hardness (and salinity in the case of marine tanks.)

There are many types of biological filters out there today (Fig. 57), many of which also incorporate other types of filtration. Some commonly used bio-filters include: under gravel filters, bio-wheels, trickle filters, canister filters, fluidized bed filters, sponge filters, and live rock/live sand have also become popular in saltwater reef aquaria.

Every aquarium is a different environment, and has it's own set of

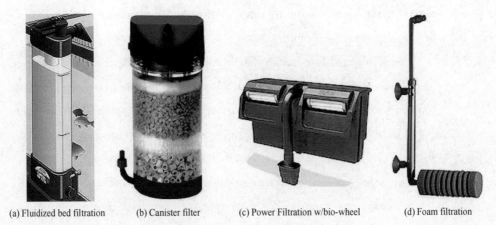

(a) Fluidized bed filtration　　(b) Canister filter　　(c) Power Filtration w/bio-wheel　　(d) Foam filtration

Fig. 57　Types of biological filters

demands. Some filters are better suited for a some aquariums than others. Consult us, or read up on what biological filtration is right for your aquarium system.

6.6.2　Key points for operation

 a. Logical filtration cycling process.

 b. System maintenance.

Water changes and filter maintenance will both affect the biological filtration to some degree. When performing water changes, it is important that the replacement water is free of any toxic chemicals such as chlorine. These chemicals can kill bacteria within the system and any water that is to be used, should be treated either by reverse osmosis, or by one of the many available liquid dechlorinators. Filter maintenance, if not done properly, can have a large effect on the biological filtration. Again, the beneficial bacteria responsible for the nitrogen cycle, populate in the greatest numbers where the water flow and oxygen content of the water are the highest. This is typically within the filter. When performing maintenance on the filter, it is ideal to leave the biological media untouched in order to preserve the bacteria. If there is no biological media within the filter, it is wise to change only of the mechanical media at a time. The remaining media that is to be reused should be rinsed in water taken from the aquarium in order to preserve the bacteria colony.

c. Restoring the balance.

When performing water changes, it is important to change no more than 25% of the aquarium water at a time. Changing more than 25% of the aquarium water can cause rapid changes in both temperature and pH, which can result in added stress to the aquarium inhabitants. Therefore, if toxins are present, it is best to perform small water changes frequently (even daily) rather than performing large water changes at less frequent intervals. Again, the makeup water that is used to replace the aquarium water should be treated by reverse osmosis, distillation, or at the very least using a liquid dechlorinator. It is ideal that the makeup water is at the same temperature as the aquarium, and has been aerated prior to adding it to the aquarium.

There are many chemical medias available on the market that will help control sudden increases in ammonia. By stopping the ammonia prior to it being broken down by the bacteria, we are reducing the biological load on the system. These products can be useful in the situations that have been described above. Again, it is important when using these products to monitor the water quality, and to perform water changes when any toxin levels are detected.

6.6.3 Outside operation practice

(1) Biological aquarium filtration

In many ways, filtration is the most complicated aspect of fish keeping. Fig. 58 shows the biological aquarium filtration.

(2) Setting up a biological filter

One way to introduce bacteria into the tank is to use some gravel or sand from an established tank when setting up your new tank. You can also buy liquid additives from the pet store that will introduce live bacteria into your tank. Once inside your tank, make sure the bacteria have a place to colonize. Sand, gravel and plant leaves all make ideal homes for a growing bacteria colony. Your power filter may have a sponge inside

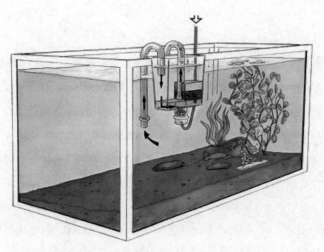

Fig. 58 Biological aquarium filtration

that's designed to house bacteria.

(3) Cleaning your filter

Every now and then you'll need to clean your filter. Care should be taken when cleaning the sponge in your filter not to wipe out a beneficial bacteria colony. Don't run your sponge under hot tap water, as this will be a death sentence for any bacteria living there. Tap water contains chlorine, which kills bacteria. It can take up to a month for a new colony of bacteria to grow. Rinse your filter in water taken out of the tank to keep your filter clean and your bacteria happy. Always use a product that removes chlorine on any tap water you put in your tank. This will keep not only your fish but your biological filter healthy.

Chapter 7 The sewers

In older cities, the sewers carry both wastewater and rainwater that run off of the streets. These are called combined sewers. During heavy rain events, the flow in combined sewers can reach capacity and threaten to back up into your house. To prevent this, relief structures called combined sewer overflows are constructed sending the excess water into streams, lakes and harbors. Combined sewer overflows can grossly pollute rivers and lakes and thus efforts are being made to limit their discharge. Today, separate sewers are installed, with rain runoff discharge to surface waters via storm sewers and wastewater routed to the treatment plant with no discharge en route.

7.1 Combined sewers

7.1.1 Fundamentals

A combined sewer is a sewage collection system of pipes and tunnels designed to also collect surface runoff (Fig. 59). This type of gravity sewer design is no longer used in building new communities (because current design separates sanitary sewers from runoff), but many older cities continue to operate combined sewers.

Combined sewers can cause serious water pollution problems during combined sewer overflow events when wet weather flows exceed the sewage treatment plant capacity. The discharges contain human and industrial waste, and can cause beach closings, restrictions on shellfish consumption and contamination of drinking water sources.

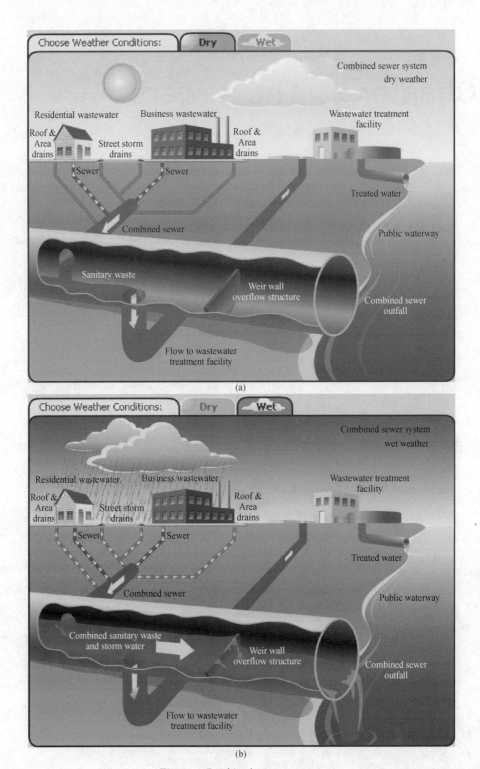

Fig. 59 Combined sewer system

7.1.2 Key points for operation

(1) Combined sewer overflows

Combined sewer overflows are constructed in combined sewer systems to divert flows in excess of the peak design flow of the sewage treatment plant. Combined sewers are built with control sections establishing stage-discharge or pressure differential-discharge relationships which may be either predicted or calibrated to divert flows in excess of sewage treatment plant capacity. A leaping weir may be used as a regulating device allowing typical dry-weather sewage flow rates to fall into an interceptor sewer to the sewage treatment plant, but causing a major portion of higher flow rates to leap over the interceptor into the diversion outfall. Alternatively, an orifice may be sized to accept the sewage treatment plant design capacity and cause excess flow to accumulate above the orifice until it overtops a side-overflow weir to the diversion outfall.

(2) Comparison to sanitary sewer overflows

Combined sewer overflows should not be confused with sanitary sewer overflows. Sanitary sewer overflows are caused by sewer system obstructions, damage, or flows in excess of sewer capacity (rather than treatment plant capacity). Sanitary sewer overflows may occur at any low spot in the sewer system rather than at the combined sewer overflows relief structures. Absence of a diversion outfall often causes sanitary sewer overflows to flood residential structures and/or flow over traveled road surfaces before reaching natural drainage channels. Sanitary sewer overflows may cause greater health risks and environmental damage than combined sewer overflows if they occur during dry weather when there is no precipitation runoff to dilute and flush away sewage pollutants.

7.1.3 Outside operation practice

    a. Sewer separation;

    b. Combined sewer overflow storage;

    c. Expanding sewage treatment capacity;

d. Retention basins;

e. Screening and disinfection facilities.

7.2　Separate sewers

7.2.1　Fundamentals

Separate sewerage consists in the separate collection of municipal wastewater (blackwater from toilet, grey water and industrial wastewater) and surface run-off (rainwater and stormwater). The separate collection prevent the overflow of sewer systems and treatment stations during rainy periods and the mixing of the relatively little polluted surface run-off with chemical and microbial pollutants from the municipal wastewater. The design of the sewers and the (semi-) centralized treatment stations thus needs to consider the volume of the wastewater only and the surface run-off and rainwater can be reused (e.g. for landscaping or agriculture) after a simplified treatment.

By replacing the combined sewer systems with separate sewer systems, sewage and surface run-off can be managed in two separate systems. The main reason for this is that surface run-off is generally less polluted than wastewater, and that treatment of combined wastewater and surface run-off is difficult during heavy rainfalls, resulting in untreated overflows. Controlling the surface run-off separately and avoiding combined sewer overflow, residents in low-lying areas in particular will avoid having their basements and ground floors flooded during extreme rain events. Should a rain event lead to flooding nonetheless, rainwater and not unsanitary sewage from kitchens and bathrooms will rise up into their basements.

Separation also eliminates the risk of sewage getting into the environment. Instead, it will be carried on to the treatment plants via a closed system, while storm water can be led to detention basins and watercourses. However, in practice there remains risk ingress of surface run-off into wastewater sewerage pipes, because of unsealed pipe joints, and unintentional or illegal connections of rainwater run-off. Conversely there may be unintentional or illegal wastewater connections to storm, water sewerage.

Fig. 60 shows separated sewer system.

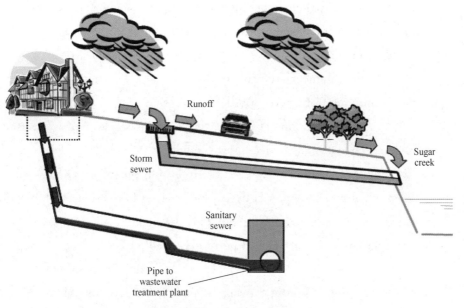

Fig. 60   Separated sewer system

7.2.2 Key point for operation

(1) Costs considerations

The construction costs might be higher than for the combined sewer system because two separated networks are necessary. These must also be maintained and operated separately. Also to replace the combined system by a separated system is very costly.

(2) Operation and maintenance

Operation and maintenance is an essential part of wastewater management and affects technology selection. Many wastewater treatment projects fail or perform poorly after construction because of inadequate operation and maintenance. On an annual basis, the operation and maintenance expenditures of treatment and sewage collection are typically in the same order of magnitude as the depreciation on the capital investment. Operation and maintenance requires:

a. Careful exhaustive planning (specialised engineers);

b. Qualified and trained staff devoted to its assignment;

c. An extensive and operational system providing spare parts and operation and maintenance utilities;

d. A maintenance and repair schedule, crew and facility;

e. A management atmosphere that aims at ensuring a reliable service with a minimum of interruptions;

f. A substantial annual budget that is uniquely devoted to operation and maintenance and service improvement.

Maintenance policy can be corrective, i. e. repair or action is undertaken when breakdown is noticed, but this leads to service interruption and hence dissatisfied customers. Ideally, maintenance is preventive, i. e. replacement of mechanical parts is carried out at the end of their expected lifetime. This allows optimal budgeting and maintenance schedules that have minimal impact on service quality. Clearly, operation and maintenance requirements are important factors when selecting a technology; process design should provide for optimal, but low cost, operation and maintenance.

7.2.3 Outside operation practice

(1) Health aspects

Already the combined sewer system technology provides a high level of hygiene and comfort. A properly constructed separated system makes it even more secure, because the sewage is transported in a closed system directly to the treatment plant and cannot overflow into the environment.

(2) Applicability

As for any sewer system, sufficient water needs to be available to carry the waste material in the sewer and therefore such systems are only applicable where enough water is available. This system is suitable for urban areas that have the resources to implement, operate and maintain such systems plus provide adequate treatment to avoid pollution at the discharge end. Separated sewers are especially suitable in areas where irregular, heavy rainfall is expected (e. g. monsoon). During these periods stormwater runoff can be managed separately and maybe stored for the dry season. Planning, construction, operation and maintenance require expert knowledge: a professional management system must be in place. When new systems are built, it is preferable to construct the separated system to prevent combined sewer overflow, limit the amount of highly-polluted wastewater and allow the reuse of surface water run-off.

Chapter 8　Characteristics of wastewater

Municipal wastewater is mainly comprised of water (99.9%) together with relatively small concentrations of suspended and dissolved organic and inorganic solids.

8.1　Organic substances

8.1.1　Fundamentals

Organic substances present in sewage include carbohydrates, lignin, fats, soaps, synthetic detergents, proteins and their decomposition products, as well as various natural and synthetic organic chemicals from the process industries. Table 10 shows the average composition of wastewater in Amman, Jordan.

Table 10　Average composition of wastewater in Amman, Jordan

Constituent	Concentration/(mg/L)
Dissolved solids(TDS)	1170
Suspended solids	900
Nitrogen(as N)	150
Phosphorus(as P)	25
Alkalinity(as $CaCO_3$)	850
Sulphate(as SO_4)	90
BOD_5	770
COD[1]	1830
TOC[2]	220

[1] COD is chemical oxygen demand.
[2] TOC is total organic carbon.
Source: Al-Salem(1987).

In arid and semi-arid countries, water use is often fairly low and sewage tends to be very strong.

8.1.2 Key points for social research

a. Definition of organic substances.

b. Classification

-Natural compounds: refer to those that are produced by plants or animals. Many of these are still extracted from natural sources because they would be more expensive to produce artificially. Examples include most sugars, some alkaloids and terpenoids, certain nutrients such as vitamin B_{12}, and, in general, those natural products with large or stereoisometrically complicated molecules present in reasonable concentrations in living organisms.

Further compounds of prime importance in biochemistry are antigens, carbohydrates, enzymes, hormones, lipids and fatty acids, neurotransmitters, nucleic acids, proteins, peptides and amino acids, lectins, vitamins, and fats and oils.

-Synthetic compounds: compounds that are prepared by reaction of other compounds are known as "synthetic". They may be either compounds that already are found in plants or animals or those that do not occur naturally. Most polymers (a category that includes all plastics and rubbers), are organic synthetic or semi-synthetic compounds.

8.1.3 Outside social research practice

a. Types of organic substance;

b. Organic substance decomposition.

8.2 Inorganic substances

8.2.1 Fundamentals

Municipal wastewater also contains a variety of inorganic substances from domestic and industrial sources, including a number of potentially toxic elements such as arsenic, cadmium, chromium, copper, lead, mercury, zinc, etc. Even if toxic materials are not present in concentrations likely to affect humans, they might well be at phytotoxic levels, which would limit their agricultural use. However, from the point of view of health, a very important consideration in agricultural use of wastewater, the contaminants of the greatest concern are the pathogenic micro-organisms and macro-organisms.

8.2.2 Key points for social research

a. Definition of inorganic substances.

b. Classification

-Simple substances: molecules consist of one-type atoms (atoms of one element). In chemical reactions cannot be decomposed with forming other substances.

-Complex substances (or chemical compounds): molecules consist of different type atoms (atoms of different chemical elements). In chemical reactions are decomposed with formation several other substances.

8.2.3 Outside social research practice

a. Types of organic substance;

b. Genetic relationship between different classes of inorganic substances (bases, oxides, acids, salts, medium salts, acid salts, basic salts and complex salts).

8.3 Pathogenic microorganisms

8.3.1 Fundamentals

Pathogenic viruses, bacteria, protozoa and helminths may be present in raw municipal.

Wastewater at the levels indicated and will survive in the environment for long periods. Pathogenic bacteria will be present in wastewater at much lower levels than the coliform group of bacteria, which are much easier to identify and enumerate (as total coliforms/100 mL). *Escherichia coli* are the most widely adopted indicator of faecal pollution and they can also be isolated and identified fairly simply, with their numbers usually being given in the form of faecal coliforms (FC) /100 mL of wastewater.

8.3.2 Key points for social research

(1) Types of pathogens

Natural or human-triggered changes in the environment might upset the natural balance between living organisms. These new environmental conditions may encourage pathogens, allowing them to multiply rapidly and increase the risk of exposing humans who share that environment. Here are the main groups of human pathogens with some examples of the diseases they cause.

a. Bacteria: $E.\,coli$ causes food poisoning and urinary tract infections. Mycobacterium tuberculosis causes tuberculosis.

b. Viruses: influenza virus causes 'flu'. Herpes simplex virus causes herpes.

c. Protozoa: plasmodium causes malaria.

d. Fungi: tinea causes ringworm.

(2) Prevention of infections

The best way of fighting disease is to prevent it. The discovery and manufacture of antibiotics and vaccines in the early part of the 20th century have changed the state of people's health forever. There are three stages of prevention of infections: primary, secondary and tertiary.

a. The Primary stage: involves public education about infectious diseases.

b. The Secondary stage: involves treating the actual infection that has already occurred by quarantining and/or vaccinating of infected individuals.

c. The Tertiary stage: involves the recovery from illness.

8.3.3 Outside social research practice

a. Identification of pathogenic microorganisms;

b. Genera and microscopy features of pathogenic microorganisms;

c. Species and clinical characteristics of pathogenic microorganisms.

Chapter 9　Preliminary wastewater treatment systems

Preliminary wastewater treatment systems include screening, grit clarifier.

9.1　Screening

9.1.1　Fundamentals

Screening is the first unit operation generally encountered in wastewater treatment plants. A screen is a device with openings, generally of uniform size used to retain solids found in the influent wastewater to the treatment plant.

The principal role of screening is to remove coarse materials from the inflow wastewater that could:

a. Damage followed treatment process equipment;

b. Reduce overall treatment process reliability and effectiveness;

c. Contaminate waste way.

There are two types of screening processes (Fig. 61):

a. Manually operated process;

b. Automatically operated process;

c. Course screens (Bar Racks): openings of 6-150mm;

d. Fine screens: openings of less than 6mm;

e. Micro screens: openings of less than 0.5μm.

9.1.2　Key points for operation

a. Design of screening chamber.

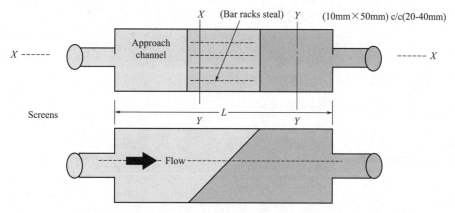

Fig. 61 Screening processes

The objective of screens is to remove large floating materials and coarse solids from wastewater. It may consist of parallel bars, wires or grating placed across the flow inclined at 30°-60°. According to method of cleaning, the screens are hand cleaned screens or mechanically cleaned screens. Whereas, according to the size of openings, the screens are classified into coarse screens ($\geqslant$50mm), medium screens (25-50mm) and fine screens (10-25mm). Normally, medium screens are used in domestic wastewater treatment.

b. Performance and technical features.

9.1.3 Outside operation practice

a. Description and working principle;

b. Design features;

c. Accessories.

9.2 Flow equalization tank

9.2.1 Fundamentals

Flow equalization is method used to overcome the operational problems and flow rate variations to improve the performance of downstream processes and to reduce the size & cost of downstream treatment facilities. To prevent flow rate, temperature, and contaminant concentrations from varying widely, flow equalization is often used.

It achieves its objective by providing storage to hold water when it is arriving too rapidly, and to supply additional water when it is arriving less rapidly than desired.

Fig. 62 shows the flow equalization.

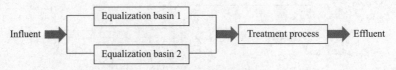

Fig. 62　Flow equalization

9.2.2　Key points for operation

a. Items we can include in an equalization tank:

-Aeration systems;

-Decanting facilities;

-Flow boxes to equalize flow over a 24 hour period of time;

-Covered or open;

-Plain concrete walls or decorative finish;

-Unique "Positive odor control system".

b. Equalization+steady state conditions=improved effluent+reduction in operating costs.

9.2.3　Outside operation practice

a. Equalization tank volume;

b. System controls.

9.3　Grit clarifier

9.3.1　Fundamentals

Grit clarifier is designed to remove grit, consisting of sand, gravel, sanders or other heavy solid materials that have specific gravities or setting velocities substantially greater than those of organic particles in wastewater. Grit chamber is generally positioned after screens and before primary sedimentation (Fig. 63).

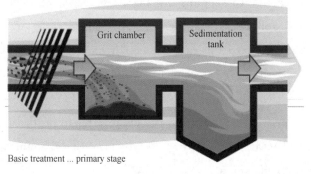

Fig. 63　Grit clarifier

Like sedimentation tanks, clarifiers are designed mainly to remove heavier particles or coarse inert and relatively dry suspended solids from the wastewater. There are two main types of grit clarifiers like rectangular horizontal flow types and aerated grit chambers. In the aerated grit chamber the organic solids are kept in suspension by rising aerated system provided at the bottom of the tank.

9.3.2　Key points for operation

(1) Purpose of grit clarifier

Grit chambers are provided to:

a. Protect moving mechanical equipment from abrasion and accompanying abnormal wear;

b. Reduce formation of heavy deposits in pipelines, channels and conduits;

c. Reduce the frequency of digester.

(2) Types of grit clarifier

a. Horizontal flow (rectangular or square, configuration type);

b. Aerated grit clarifier;

c. Vortex type grit clarifier.

9.3.3　Outside operation practice

The grit channel is a rectangular chamber that slows the speed of flow and allows inorganic materials (minerals, sand, gravel, egg shells, glass, coffee grounds, etc.) to fall to the bottom of the chamber. The flow is still fast enough to keep organic materials (decayed food, shredded toilet tissue, toilet wastes, sugars, fats, and grease) in suspension and assures it floats to the next process which is primary treatment. Grit buckets are connected to chains that run along the bottom of the chamber and scoop the grit from the bottom of the tank.

The grit is removed from the wastewater to protect pumps from abrasion and wear. Grit removal also reduces the clogging in the pipes to the primary clarifiers.

Chapter 10 Primary wastewater treatment systems

Sedimentation or setting tanks that receive raw wastewater prior to biological treatment are called primary tanks. The objective of the primary sedimentation tank (Fig. 64) is to remove readily settleable organic solids and floating materials and thus reduce the suspended solids content. Efficiently designed and operated primary sedimentation tanks should remove from 50% to 70% the suspended solids and 25% to 40% of the BOD.

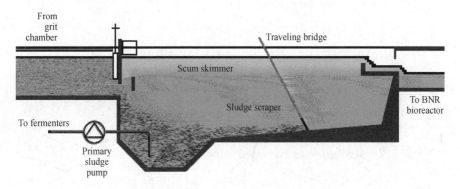

Fig. 64 Primary sedimentation tank

10.1 Plain sedimentation

10.1.1 Fundamentals

Primary sedimentation is generally plain sedimentation without the use of chemicals.

a. Circular sedimentation tank.

Most commonly have diameters from 3 m to 60 m (side water depth range from 3 m to 5 m) (Fig. 65).

Fig. 65 Circular sedimentation tank

b. Rectangular sedimentation tank.

Length ranges of 15 m to 100 m and width from 3 m to 24 m (length/width ratio 3∶1 to 5∶1).

c. Square sedimentation tank.

The sedimentation tanks may be flat bottomed or hopper bottomed. Wastewater enters the tanks, usually at the center, through a well or diffusion box. In the quiescent period, the suspended particles settle to the bottom as sludge and are raked towards a central hopper from where the sludge is withdrawn.

10.1.2 Key points for operation

Traditionally the design criteria for sizing setting tanks are:

a. Average overflow rate: 30-50 $m^3/(m^2 \cdot d)$ [typical 40 $m^3/(m^2 \cdot d)$].

b. Peak hourly overflow rate: 50-120 $m^3/(m^2 \cdot d)$.

c. Weir loading rate: 1.5-2.5 h (typical 2.0 h).

The settling tank design in such cases depends on both surface loading and detention time.

Fig. 66 shows the screen & aerated grit chamber.

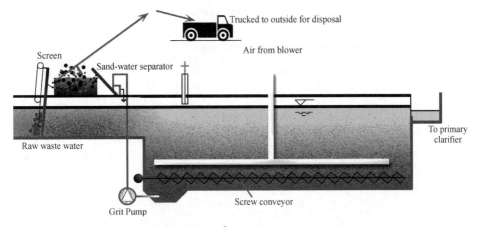

Fig. 66 Screen & Aerated grit chamber

10.2 Aerated sedimentation

10.2.1 Fundamentals

Aerated sedimentation is to conduct aeration within sedimentation tank to control odor and smells, and the same to achieve a certain removal for organic substances, thus reduce the BOD loading to the followed biological treatment in the secondary treatment process. (Fig. 67)

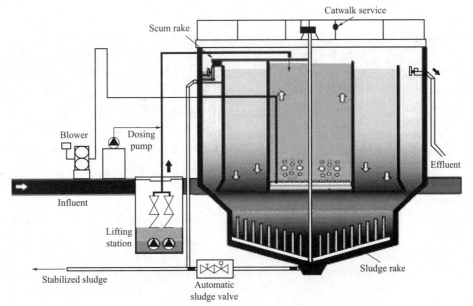

Fig. 67 Aerated sedimentation

10.2.2 Outside operation practice

a. Design of aeration sedimentation;

b. Assessment of main process characteristics;

c. Settling of discrete particles;

d. Settling of flocculent particles;

e. Zone-settling behaviour.

Chapter 11 Secondary wastewater treatment systems

Secondary treatment is typically biological oxidation to remove biochemical oxygen demand by removal of dissolved and fine suspended biodegradable organic matter from wastewater. Secondary wastewater treatment systems usually utilize biological treatment processes (Fig. 68). There are several different approaches to biological wastewater treatment. The most commonly used are activated sludge, trickling filter and oxidation ditch treatment systems. The key is to bring aerobic microorganisms, organic matter in wastewater, and dissolved oxygen together.

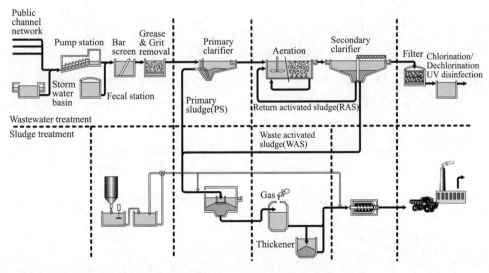

Fig. 68 Secondary wastewater treatment systems

11.1 Conventional activated sludge process

11.1.1 Fundamentals

The most common secondary wastewater treatment process used for municipal wastewater treatment is the activated sludge process (Fig. 69).

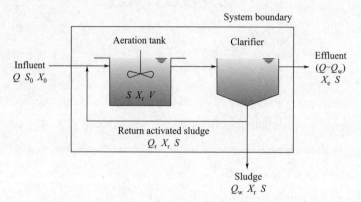

Fig. 69 Conventional activated sludge process

(1) Activated sludge plant involves

a. Wastewater aeration in the presence of a microbial suspension;

b. Solid-liquid separation following aeration;

c. Discharge of clarified effluent;

d. Wasting of excess biomass;

e. Return of remaining biomass to the aeration tank.

In this process, wastewater after primary treatment is aerated in an aeration basin in which micro-organisms metabolize the suspended and soluble organic matter:

a. About 30% of the organic matter is oxidized to CO_2 and water to derive energy;

b. The remaining 70% is synthesized into new cells.

The new cells formed in the reaction are removed from the liquid in the form of a flocculent sludge in settling tanks:

a. A part of this settled activated sludge is returned to the aeration tank to maintain expected higher cell concentration;

b. The remaining part is referred as waste or excess sludge for sludge treatment.

(2) Activated sludge process variables

The main variables of activated sludge process are the mixing regime, loading rate, and the flow scheme.

a. Mixing regime. Generally two types of mixing regimes are of major interest in activated sludge process: plug flow and complete mixing. In the first one, the regime is characterized by orderly flow of mixed liquor through the aeration tank with no element of mixed liquor overtaking or mixing with any other element. There may be lateral mixing of mixed liquor but there must be no mixing along the path of flow.

In complete mixing, the contents of aeration tank are well stirred and uniform throughout. Thus, at steady state, the effluent from the aeration tank has the same composition as the aeration tank contents.

The type of mixing regime is very important as it affects oxygen transfer requirements in the aeration tank, susceptibility of biomass to shock loads, local environmental conditions in the aeration tank, and the kinetics governing the treatment process.

b. Loading Rate. Loading parameters that have been developed mainly includes:

Hydraulic retention time (HRT), t (day):
$$t = V/Q$$
$V =$ volume of aeration tank, m^3, and $Q =$ sewage inflow, m^3/d.

Volumetric organic loading: an empirical loading parameter, which is defined as the BOD applied per unit volume of aeration tank, per day.

A rational loading parameter which has found wider acceptance and is preferred is specific substrate utilization rate, L_s, per day.
$$L_s = Q(S_0 - S_e)/(V\ X)$$

A similar loading parameter is mean cell residence time or sludge retention time, SRT.
$$SRT = V\ X/[Q_w X_r + (Q - Q_w) X_e]$$

where, S_0 and S_e are influent and effluent organic matter concentration respectively, measured as BOD_5 (g/m^3); X, X_e and X_r are MLSS concentration in aeration tank, effluent and return sludge respectively, and Q_w is waste activated sludge rate.

Under steady state operation, the mass of waste activated sludge is given by:
$$Q_w X_r = YQ(S_0 - S_e) - k_d XV$$
where, Y is maximum yield coefficient (microbial mass synthesized/mass of substrate utilized) and k_d is endogenous decay rate (d^{-1}).

If the value of S_e is small as compared S_0, the loading rate may also be expressed as food to microorganism ratio (F/M):
$$F/M = Q(S_0 - S_e)/(XV) = QS_0/(XV)$$
The value of SRT adopted for design controls the effluent quality, and settleability and drainability of biomass, oxygen requirement and quantity of waste activated sludge.

c. Flow Scheme.

The flow scheme involves:

-The pattern of sewage addition;

-The pattern of sludge return to the aeration tank;

-The pattern of aeration.

Sewage addition may be at a single point at the inlet end or it may be at several points along the aeration tank. The sludge return may be directly from the settling tank to the aeration tank or through a sludge re-aeration tank. Aeration may be at a uniform rate or it may be varied from the head of the aeration tank to its end.

11.1.2 Key points for operation

Conventional system and its modifications. The conventional system maintains a plug flow hydraulic regime. several modifications to the conventional system have been developed to meet specific treatment objectives.

a. Step aeration: wastewater after primary treatment is introduced at several points along the aeration tank length which produces more uniform oxygen demand throughout.

b. Tapered aeration: attempts to supply air to match oxygen demand along the length of the tank.

c. Contact stabilization: provides for reaeration of return activated sludge from the final clarifier, which allows a smaller aeration or contact tank.

d. Completely mixed process: aims at instantaneous mixing of the influent waste and return sludge with the entire contents of the aeration tank.

e. Extended aeration process: operates at a low organic load producing lesser quantity of well stabilized sludge.

11.1.3 Outside operation practice

a. Operation and maintenance of activated sludge process;

b. Monitoring, process control, and troubleshooting;

c. Safety and regulations.

11.2 Oxidation ditch process

11.2.1 Fundamentals

An oxidation ditch refers to a modified activated sludge biological treatment process utilizing long solids retention times (SRT) to remove biodegradable organics. Although considered as typically complete mix systems, oxidation ditches can be modified through the provision of diffused air to approach plug flow conditions (Fig. 70).

Typical oxidation ditch treatment systems consist of a single or multichannel configuration within a ring, oval, or horseshoe-shaped basin, with the provision of horizontally-mounted or vertically-mounted aerators. These aerators are responsible for facilitating circulation, oxygen transfer, and aeration in the ditch.

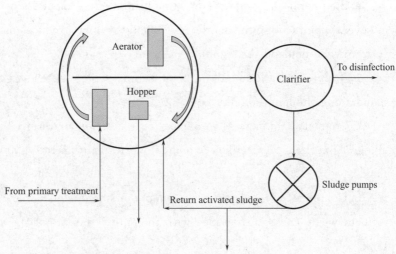

Source: Parsons Engineering Science, Inc. 2000.

Fig. 70 Oxidation ditch process

11.2.2 Key points for operation

(1) Applications

a. The oxidation ditch process is suitable in any situation where activated sludge treatment (conventional or extended aeration) is appropriate.

b. The ditches are also applicable in plants requiring enhanced nitrification.

c. This process, requiring more land compared with conventional treatment facilities, is shown to be highly effective in small installations, small communities, and isolated institutions.

(2) Major advantages

a. The constant water level and continuous discharge, which lowers the weir overflow rate and eliminates the periodic effluent surge, make the technology reliable over other biological processes.

b. Its long hydraulic retention time and complete mixing reduces the impact of a shock load or hydraulic surge.

c. Because of its extended biological activity during the activated sludge process, the oxidation ditch produces less sludge compared with other biological treatment processes.

d. The process is energy-efficient.

(3) Some disadvantages

a. Effluent suspended solid concentrations are relatively high compared to other modifications of the activated sludge process.

b. The process requires a larger land area.

11.2.3 Outside operation practice

a. Operation and maintenance of oxidation ditch process;

b. Monitoring, process control, and troubleshooting;

c. Safety and regulations.

11.3 Trickling filter process

11.3.1 Fundamentals

Different from activated sludge process, known as suspended-growth process, in which microorganisms are sustained in liquid; the trickling filter process is known as attached-growth process.

Trickling filter process is an aerobic treatment system that utilizes microorganisms attached to a medium to remove organic matter from wastewater (Fig. 71).

This type of system is common to a number of technologies such as rotating biological contactors and packed bed reactors.

11.3.2 Key points for operation

(1) Applications

a. Trickling filter process enables organic matter in wastewater to be adsorbed by a population of microorganisms (aerobic, anaerobic, and facultative bacteria; fungi; algae; and protozoa) attached to the medium as a biological film or slime layer (approximately 0.1-0.2 mm thick).

b. As the wastewater flows over the medium, microorganisms already in the water gradually attach themselves to the rock, slag, or plastic surface and form a film. The organic matter material is then degraded by the aerobic mi-

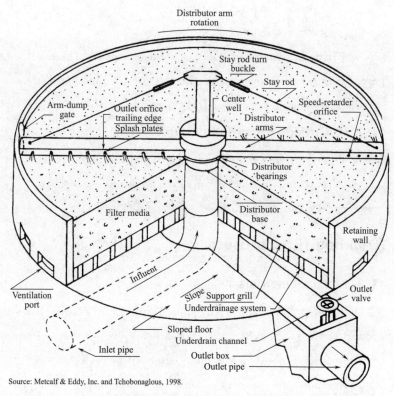

Source: Metcalf & Eddy, Inc. and Tchobonaglous, 1998.

Fig. 71 Trickling filter process

croorganisms in the outer part of the slime layer.

c. As the layer thickens through microbial growth, oxygen cannot penetrate the medium face, and anaerobic organisms develop. As the biological film continues to grow, the microorganisms near the surface lose their ability to cling to the medium, and a portion of the slime layer falls off the filter.

d. This process is known as sloughing. The sloughed solids are picked up by the underdrain system and transported to a clarifier for removal from the wastewater.

(2) Major advantages

a. Simple, reliable, biological process.

b. Suitable in areas where large tracts of land are not available for land intensive treatment systems.

c. May qualify for equivalent secondary discharge standards.

 d. Effective in treating high concentrations of organics depending on the type of medium used.

 e. Appropriate for small-sized to medium-sized communities.

 f. Rapidly reduce soluble BOD_5 in applied wastewater.

 g. Efficient nitrification units.

 h. Durable process elements.

 i. Low power requirements.

 j. Moderate level of skill and technical expertise needed to manage and operate the system.

(3) Some disadvantages

 a. Additional treatment may be needed to meet more stringent discharge standards.

 b. Possible accumulation of excess biomass that cannot retain an aerobic condition and can impair tricking filter performance (maximum biomass thickness is controlled by hydraulic dosage rate, type of media, type of organic matter, temperature and nature of the biological growth).

 c. Requires regular operator attention.

 d. Incidence of clogging is relatively high.

 e. Requires low loadings depending on the medium.

 f. Flexibility and control are limited in comparison with activated-sludge processes.

 g. Vector and odor problems.

 h. Snail problems.

11.3.3 Outside operation practice

(1) Design criteria

 a. Permeable medium: made of a bed of rock, slag, or plastic over which wastewater is distributed to trickle through.

 b. Rock or slag beds can be up to 60.96 meters in diameter and 0.9-2.4 meters in depth with rock size varying from 2.5-10.2 cm. Most rock media provide approximately 149 m^2/m^3 of surface area and less than 40 percent void space.

c. Packed plastic filters (bio-towers), on the other hand, are smaller in diameter (6-12 meters and range in depth from 4.3 to 12.2 meters.

d. The filters look more like towers, with the media in various configurations (e. g. vertical flow, cross flow, or various random packings).

e. The design of a TF system for wastewater also includes a distribution system. Rotary hydraulic distribution is usually standard for this process, but fixed nozzle distributors are also being used in square or rectangular reactors. Fixed nozzle distributors are being limited to small facilities and package plants.

f. Trickling filter has an underdrain system that collects the filtrate and solids, and also serves as a source of air for the microorganisms on the filter. The treated wastewater and solids are piped to a settling tank where the solids are separated.

g. Usually, part of the liquid from the settling chamber is recirculated to improve wetting and flushing of the filter medium, optimizing the process and increasing the removal rate.

h. It is essential that sufficient air be available for the successful operation of the process. To supply air to the system, natural draft and wind forces are usually sufficient if large enough ventilation ports are provided at the bottom of the filter and the medium has enough void area.

Table 11 shows the removal rates of BOD_5 with various filter rates.

Table 11 Removal rates of BOD_5 with various filter rates

Filter type	BOD_5 removal/%
Low rate	80-90
Intermediate rate	50-70
High rate	65-85
Roughing filter	40-65

Source: Environmental Engineers Handbook, 1997.

(2) Cautions for operation and maintenance

a. Disagreeable odors from Filter;

b. Ponding on filter media;

c. Filter flies (psychoda);

d. Icing;

e. Rotating distributor slows down or stops.

11.4 Anaerobic waste treatment processes

11.4.1 Fundamentals

Advantage of anaerobic waste treatment systems as means for recovery of non-conventional energy is increasingly being recognized worldwide.

Anaerobic decomposition is a biologically mediated process, indigenous to nature, and capable of being simulated for treating wastes emanating from municipal, agricultural, and industrial activities.

Anaerobic digestion as applied in treatment of sewage sludge and other organic wastes, represents the controlled application of a process. Anaerobic digesters have traditionally been used for many decades in the stabilization of sewage sludge, however, their successful and economic employment for the treatment of wastewater is only a recent phenomenon, arising from the development of new reactor designs. These concepts have led to development of various reactors, which are capable of retaining a much higher biomass concentration than traditional digesters. Making the sludge retention independent of the influent retention time makes this possible.

The technological approaches to allow this condition of independent sludge retention time can be divided in to the following:

a. Attachment of biomass on the media (filters, fluidized systems, and RBC configurations);

b. Non-attached biomass concept as suspended growth process (sludge blanket reactors and contact process with sludge recycling).

Table 12 shows basic types of anaerobic process reactors.

Table 12 Basic types of anaerobic process reactors

Type of reactor	Synonyms	Abbreviations
Attached Biomass		
Fixed bed	Fixed film, filter, submerged filter, stationary fixed bed	SMAR (submerged anaerobic reactor)/ANFIL(upflow anaerobic filter), AUF (anaerobic upflow filter), ADSR (anaerobic downflow stationary bed reactor), AF(anaerobic filter), DSFF (downflow stationary fixed film reactor)
Moving bed	Rotting discs, rotating biological contactor,	AnRBC, RBC
Expanded bed	Anaerobic attached film expanded bed	AAFEB
Fluidized bed	Anaerobic fluidized bed reactor, carrier-assisted contact process	FBBR/IFCR (immobilized fluidized cells reactor)/CASBER (carrier-assisted sludge bed reactor)
Non-attached biomass		
Recycled flocks, sludge blanket, digester	Contact process, upflow anaerobic sludge blanket reactor (UASB), upflow sludge blanket (USB), clarigester type	UASB, USB

Reference: Henze and Harremoes, 1983.

(1) Anaerobic filter systems

Down-flow stationary fixed film (DSFF) reactor: packed with a fixed support media. Biomass is present as a biofilm attached to the support media. Most of the biomass is present as suspended and/or entrapped biomass in the interstitial pore volume of the support media. The other major difference between these reactors is the direction of liquid flow through the packing. Different from the AF, for which, the feed enters at bottom of the reactor, in the DSFF reactor, influent is applied in downward direction from the top of the rector. Both reactors can be used to treat either diluted or concentrated soluble wastewater.

Because of the relatively large clearance between the channels in the vertically oriented media used, the DSFF reactor is able to treat wastewater with relatively high suspended solids while the AF cannot. The bacteria are retained on the media and not washed off in the effluent, thus, means cell residence times of the order of 100 days can be obtained. In the AF, most of the biological activity is due to the biomass in suspension (entrapped) rather than to the attached biofilm. The media with a high capacity to entrap and prevent washout of the biomass from the reactor is more important than the specific surface area (surface area to volume ratio) of the media. The biofilm thickness of 1-3 mm has been observed in fixed-bed reactor.

In the DSFF reactor systems, virtually all of the active biomass is attached to the support media. Different types of support media such as needle punched polyester (NPP) and red drain tile clay, PVC or glass can be used. For NPP, this attachment is probably associated with its surface roughness. The leaching of minerals from the clay could potentially stimulate bacterial activity and adhesion to this media support.

Selection of proper inoculum source is important to obtain rapid reactor start-up and minimize the time required for the initial biofilm establishment. Usually a bacterial flora adapted to the target wastewater should be used. In general, the volume of inoculum used should at least 10% (v/v) to obtained good result. During the start-up period the initial specific load applied should be maintained at levels less than 0.3 kg COD/(kg VSS · d) and hydraulic retention time (HRT) greater than 1 day should be maintained to prevent wash out of the inoculated biomass.

Typical organic loading rates generally between 1.0 kg COD/(m^3 · d) and 10 kg COD/(m^3 · d) can be applied with 75%-85% removal efficiency. HRT is generally kept in the range of 18-24 h, but lower HRT values can also give fairly good removal efficiency depending on the type of organic matter present in the wastewater.

(2) Expanded bed process

In the expanded bed process, the wastewater to be treated is pumped upward through a bed of appropriate medium (e. g. sand, coal, expanded aggregate, plastic media) on which a biological growth has been developed. Effluent is recycled to dilute the incoming wastewater and to provide an adequate flow to maintain the bed in an expanded condition.

Biomass concentrations exceeding 15000-40000 mg/L can be developed. Since more biomass can be maintained, the expanded bed process can also be used for the treatment of low strength wastewater, such as municipal sewage, at very short HRTs. Organic loading in the range of 5-10 kg COD/(m^3 · d) can be applied with COD removal efficiency of 80%-85%. HRT is generally in the range of 5-10 h.

(3) Anaerobic Contact Process

The essential feature of the anaerobic contact process is that the washout of the active anaerobic bacterial mass from the reactor is controlled by a sludge separation and recycles system.

The major problem in the practical application of the contact process has always been the separation (and concentration) of the sludge from the effluent solution. For this purpose, several methods have been used or were recommended for use, e. g. plain sedimentation, settling combined with chemical flocculation, with vacuum degasification, floatation and centrifugation.

A basic idea underlying the contact process is that it is considered necessary to thoroughly mix the digester contents, e. g. by gas recirculation, sludge recirculation, or continuous or intermittent mechanical agitation. This is generally used for concentrated wastewater treatment such as distillery wastewater.

(4) Upflow anaerobic sludge blanket (UASB) reactor

This reactor is a modified version of the contact process, based on an upward movement of the liquid waste through a dense blanket of anaerobic sludge (Fig. 72).

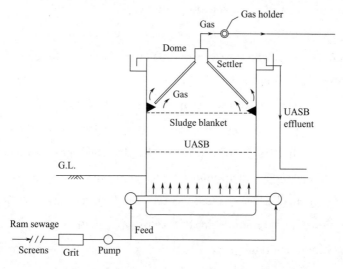

Fig. 72 Upflow anaerobic sludge blanket reactor

No inert medium is provided in the systems. The biomass growth takes place on the fine sludge particles, which then develop as sludge granules of high specific gravity.

The reactor can be divided in three parts: sludge bed, sludge blanket and three phase separator (gas-liquid-solid, GLS separator) provided at the top of the reactor.

The sludge bed consists of high concentration of active anaerobic bacteria (40-100 g/L) and it occupies about 40%-60% of reactor volume. Majority of organic matter degradation ($>95\%$) takes place in this zone. The sludge consists of biologically formed granules or thick flocculent sludge. Treatment occurs as the wastewater comes in contact with the granules and/or thick flocculent sludge.

The gases produced causes internal mixing in the reactor. Some of the gas produced within the sludge bed gets attached to the biological granules. The free gas and the particles with the attached gas rise to the top of the reactor.

On the top of sludge bed and below GLS separator, thin concentration of sludge is maintained, which is called as sludge blanket. This zone occupies 15%-25% of reactor volume.

Maintaining sludge blanket zone is important to dilute and further treat the wastewater stream that has bypassed the sludge bed portion following the rising biogas.

The GLS separator occupies about 20%-30% of the reactor volume. The particles that raise to the liquid surface strike the bottom of the degassing baffles, which causes the attached gas bubbles to be released. The degassed granules typically drop back to the surface of the sludge bed. The free gas and gas released from the granules is captured in the gas collection domes located at the top of the reactor.

Liquid containing some residual solids and biological granules passes into a settling chamber, where the residual solids are separated from the liquid. The separated solids fall back through the baffle system to the top of the sludge blanket.

11.4.2 Outside operation practice

a. Operation and maintenance;

b. Monitoring, process control, and troubleshooting;

c. Safety and regulations.

Chapter 12   Sludge treatment and disposal

Sewage sludge treatment describes the processes used to manage and dispose of sewage sludge produced during sewage treatment. Primary sludge includes settleable solids removed during primary treatment in primary clarifiers. Secondary sludge separated in secondary clarifiers includes treated sewage sludge from secondary treatment bioreactors.

Sludge treatment is focused on reducing sludge weight and volume to reduce disposal costs, and on reducing potential health risks of disposal options. Water removal is the primary means of weight and volume reduction, while pathogen destruction is frequently accomplished through heating during thermophilic digestion, composting, or incineration. The choice of a sludge treatment method depends on the volume of sludge generated, and comparison of treatment costs required for available disposal options. Air-drying and composting may be attractive to rural communities, while limited land availability may make aerobic digestion and mechanical dewatering preferable for cities, and economies of scale may encourage energy recovery alternatives in metropolitan areas.

Energy may be recovered from sludge through methane gas production during anaerobic digestion or through incineration of dried sludge, but energy yield is often insufficient to evaporate sludge water content or to power blowers, pumps, or centrifuges required for dewatering. Coarse primary solids and secondary sewage sludge may include toxic chemicals removed from liquid sewage by sorption onto solid particles in clarifier sludge. Reducing sludge volume may increase the concentration of some of these toxic chemicals in the sludge.

12.1　Characteristics of sludge

12.1.1　Fundamentals

By analyzing the different characteristics of the activated sludge or the sludge quality, plant operators are able to monitor how effective the treatment plant's process is. Efficient operation is ensured by keeping accurate, up-to-date records; routinely evaluating operating and laboratory data; and troubleshooting, to solve problems before they become serious. A wide range of laboratory and visual and physical test methods are recommended. Principally, these include floc and settleability performance using a jar test, microscopic identification of the predominant types of bacteria, and analysis of various chemical parameters.

The treatment environment directly affects microorganisms. Changes in food, dissolved oxygen, temperature, pH, total dissolved solids, sludge age, presence of toxins, and other factors create a dynamic environment for the treatment organisms. The operators can change the environment (the process) to encourage or discourage the growth of specific microorganisms.

Table 13 shows the characteristics of the activated sludge or sludge quality, as well as plant operators.

Table 13　Characteristics of the activated sludge or sludge quality

Problem	Effect/Observation
Poor primary clarification	Plugging
	Standing water
	Odors
	Reduced efficiency
Hydraulic overload	High effluent TSS
Nitrification	High effluent TSS
	High chlorine demand
	Low pH

Continued Table 13

Problem	Effect/Observation
Nutrient shortage	Filamentous bacteria
	Rising sludge
	Pass through of soluble BOD
Organic overload	Pass through of soluble BOD
	Odors
	Low DO
	Poor effluent quality
Cold weather	Loss in removal efficiency
	Icing problems
Organic underload	High energy use
	Nitrification

12.1.2　Key points for operation

Table 14 shows the sources of solids from conventional wastewater treatment plants.

Table 14　Sources of solids from conventional wastewater treatment plants

Unit operation or process	Type of solids	Remarks
Screening	Coarse solids	Coarse solids are removed by mechanical and hand-cleaned bar screens. In small plants, screenings are often comminuted for removal in subsequent treatment units
Grit removal	Grit and scum	Scum-removal facilities are often omitted in grit-removal facilities
Pre-aeration	Grit and scum	In some plants, scum-removal facilities are not provided in pre-aeration tanks. If the pre-aeration tanks are not preceded by grit-removal facilities, grit deposition may occur in pre-aeration tanks
Primary sedimentation	Primary solids and scum	Quantities of solids and scum depend upon the nature of the collection system and whether industrial wastes are discharged to the system

Continued Table 14

Unit operation or process	Type of solids	Remarks
Biological treatment	Suspended solids	Suspended solids are produced by the biological conversion of BOD. Some form of thickening may be required to concentrate the waste sludge stream from the biological treatment system
Secondary sedimentation	Secondary biosolids and scum	Provision for scum removal from secondary settling tanks is a requirement of the U. S. EPA
Solids processing facilities	Solids, compost and ashes	The characteristics of the end products depend on the characteristics of the solids treated and operations and processes used. Regulations for the disposal of residuals are stringent

Conventional characterization parameters can be grouped in physical, chemical and biological parameters.

a. Physical parameters give general information on sludge processability and handlability;

b. Chemical parameters are relevant to the presence of nutrients and toxic/dangerous compounds, so they become necessary in the case of utilization in agriculture;

c. Biological parameters give information on microbial activity and organic matter/pathogens presence, thus allowing the safety of use to be evaluated.

12.1.3 Outside operation practice

(1) Settling characteristics

Sludge can be characterized by how well it settles. Most settling tests are conducted in a 1L graduated cylinder. A quick and simple test to measure the sludge settle ability is called the sludge volume index (SVI).

SVI is conducted with a homogeneous sludge mixture. The sludge is settled out in 1L cone for 30 minutes. Settled sludge volume (V) at the bottom of the cone is measured. Knowing the concentration of solids in the sludge suspension (MLSS) in terms of mg/L, SVI is calculated as:

$$SVI = \frac{V \times 1000}{MLSS}$$

Where, typically SVI is used without a unit, but its unit is mL/g.

Typical SVI values and their meanings:

Sludge Volume Index (SVI) values: pin floc potential less than 50 mL/g;

Sludge Volume Index (SVI) values: good range 50 mL/g to 100 mL/g;

Sludge Volume Index (SVI) values: Filament growth 100 mL/g to 150 mL/g;

Sludge Volume Index (SVI) values: Bulking at high flows 150 mL/g to 200 mL/g;

Sludge Volume Index (SVI) values: Bulking 200 mL/g to 300 mL/g;

Sludge Volume Index (SVI) values: Severe bulking higher than 300 mL/g.

(2) Particle size

Measurement of particle size in sludge is a big problem. In the past, methods like filtration through a series of different sized filters, photographic techniques, scattering laser light have been all used, but each of these have their own handicaps. Fig. 73 shows the particle size of sludge.

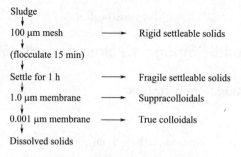

Fig. 73 Particle size of sludge

Sludge particle size affects sludge properties like dewatering, settling and rheology.

(3) Distribution of water in sludge

Sludge is a two-phase slurry, consisting of water and solids. When this itself is a problem to be handled, a more important problem is that this water in sludge is not only in one form in terms of its binding characteristics with solids.

Water in sludge appears to exist in four forms:

a. Free water: water that is not attached to sludge solids and that can be separated by simple gravitational settling. It is about 75% of the total volume.

b. Interstitial water: water that is trapped within the floc structure and travels with the floc or perhaps water trapped within a cell. This water can be released when the floc is broken up or the cell is destroyed. Some interstitial water might be removed by mechanical dewatering devices such as centrifuges. It is about 20% of the total volume.

c. Vicinal water: water that is associated with solid particles. This water is held on particle surfaces by virtue of the molecular structure of the water molecules and cannot be removed by centrifugation or other mechanical means. Vicinal water will not be free and it will exist as long as there is a surface. It is about 2% of the total volume.

d. Water of hydration: water that is chemically bound to the particle and can be released only by thermo-chemical destruction of the particles. It is about 2.5% of the total volume.

12.2 Sludge conditioning, thickening and dewatering

12.2.1 Sludge conditioning

Sludge conditioning is a process whereby sludge solids are treated with chemicals or various other means to prepare the sludge for dewatering processes, in other words, to improve dewatering characteristics of the sludge.

(1) Mechanisms of sludge conditioning

There are two mechanisms involved in sludge conditioning:

a. Neutralization of charge (double layer theory).

The double layer model is used to visualize the ionic environment in the vicinity of a charged colloid and explains how electrical repulsive forces occur. It is easier to understand this model as a sequence of steps that would take place around a single negative colloid if its neutralizing ions were suddenly stripped away. We first look at the effect of the colloid on the positive ions (often called counter-ions) in solution. Initially, attraction from the negative colloid causes some of the positive ions to form a firmly attached layer around the surface of the colloid; this layer of counter-ions is known as the stern layer. Additional positive ions are still attracted by the negative colloid, but now they are repelled by the Stern layer as well as by other positive ions that are also trying to approach the colloid. This dynamic equilibrium results in the formation of a diffuse layer of counter-ions. They have a high concentration near the surface which gradually decreases with distance, until they reachequilibrium with the counter-ion concentration in the solution. In a similar, but opposite, fashion there is a lack of negative ions in the neighborhood of the surface, because they are repelled by the negative colloid. Negative ions are called co-ions because they have the same charge as the colloid.

Their concentration will gradually increase with distance, as the repulsive forces of the colloid are screened out by the positive ions, until equilibrium is again reached. The diffuse layer can be visualized as a charged atmosphere surrounding the colloid. The charge density at any distance from the surface is equal to the difference in concentration of positive and negative ions at that point. Charge density is the greatest near the colloid and gradually diminishes toward zero as the concentration of positive and negative ions merge together. The attached counter-ions in the Stern layer and the charged atmosphere in the diffuse layer are what we refer to as the double layer. The thickness of this layer depends upon the type and concentration of ions in solution.

Fig. 74 shows the visualization of the double layer theory.

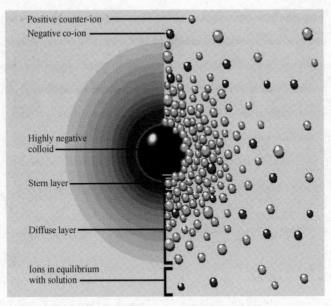

Fig. 74 Visualization of the double layer theory

b. Polymer bridge formation.

Polymers that are anionic and non-ionic (usually anionic to a slight extent when placed in water) become attached at a number of adsorption sites to the surface of the particles found in the wastewater. A bridge is formed when two or more particles become adsorbed along the length of the polymer. The size of resulting three-dimensional particles grows until they can be removed easily by sedimentation. Where particle removal is to be achieved by the formation of the particle-polymer bridges, the initial mixing of the polymer and the wastewater containing the particles to be removed must be accomplished in a matter of seconds. Instantaneous initial mixing usually not required as the polymers are already formed, which is not the case with the polymers formed by metal salts. The polymer bridge formation is shown in Fig. 75.

(2) Methods of sludge conditioning

Sludge and biosolids are generally conditioned chemically. Other conditioning methods are heat treatment and freeze-thaw methods.

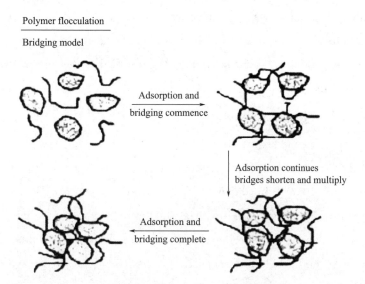

Fig. 75　Polymer bridge formation

　　a. Chemical conditioning (sludge conditioning): prepares the sludge for better and more economical treatment with vacuum filters or centrifuges. Many chemicals have been used such as sulfuric acid, alum, chlorinated copperas, ferrous sulfate, and ferric chloride with or without lime, and others.

　　Factors affecting chemical conditioning: source, solids concentration, particle size and distribution, pH and alkalinity, surface charge and degree of hydration and other physical factors.

　　b. Thermal conditioning: there are two basic processes for thermal treatment of sludge. One, wet air oxidation, is the flameless oxidation of sludge at temperatures of 450-550 °F and pressures of about 1200 psig. The other type, heat treatment, is similar but carried out at temperatures of 350-400 °F and pressures of 150-300 psig. Wet air oxidation reduces the sludge to an ash and heat treatment improves the dewaterability of the sludge. The lower temperature and pressure heat treatment is more widely used than the oxidation process.

　　c. Freeze-thaw conditioning: it is well-known fact that natural freezing of water and wastewater treatment plant residuals in cold climates enhances their dewatering characteristics. Freezing and thawing convert the jellylike consistency of the residuals to a granular-type that drains readily. Similar results have been achieved by the use of mechanical freeze/thaw equipment.

12.2.2 Sludge thickening

Thickening is often the first step in a sludge treatment process. Sludge from primary or secondary clarifiers may be stirred (often after addition of clarifying agents) to form larger, more rapidly settling aggregates. Primary sludge may be thickened to about 8% or 10% solids, while secondary sludge may be thickened to about 4% solids. Thickeners often resemble a clarifier with the addition of a stirring mechanism. Thickened sludge with less than 10% solids may receive additional sludge treatment while liquid thickener overflow is returned to the sewage treatment process. Fig. 76 shows a sewage sludge thickener.

Fig. 76 A sewage sludge thickener

12.2.3 Sludge dewatering

Water content of sludge may be reduced by centrifugation, filtration, and/or evaporation to reduce transportation costs of disposal, or to improve suitability for composting. Centrifugation may be a preliminary step to reduce sludge volume for subsequent filtration or evaporation. Filtration may occur through underdrains in a sand drying bed or as a separate mechanical process in a belt filter press. Filtrate and centrate are typically returned to the sewage treatment process. After dewatering sludge may be handled as a solid containing 50%-75% water. Dewatered sludges with higher moisture content are usually handled as liquids.

Fig. 77 shows the schematic of a belt filter press to dewater sewage sludge.

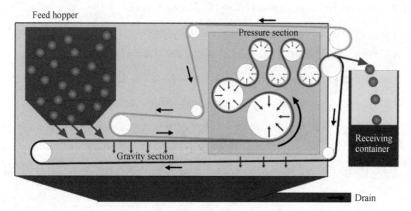

Fig. 77　Schematic of a belt filter press to dewater sewage sludge
(Filtrate is extracted initially by gravity, then by squeezing the cloth through rollers)

12.2.4　Outside operation practices

a. Zeta potential vs. double layer;

b. Operating principle;

c. Controlling municipal energy costs.

12.3　Digestion

12.3.1　Fundamentals

Many sludge are treated using a variety of digestion techniques, the purpose of which is to reduce the amount of organic matter and the number of disease-causing microorganisms present in the solids.

(1) Anaerobic digestion

Anaerobic digestion is a bacterial process that is carried out in the absence of oxygen. The process can either be thermophilic digestion, in which sludge is fermented in tanks at a temperature of 55℃, or mesophilic, at a temperature of around 36℃. Though allowing shorter retention time (and thus smaller tanks), thermophilic digestion is more expensive in terms of energy consumption for heating the sludge.

Mesophilic anaerobic digestion is also a common method for treating sludge produced at sewage treatment plants. The sludge is fed into large tanks and held for a minimum of 12 days to allow the digestion process to perform the four stages necessary to digest the sludge. These are hydrolysis, acidogenesis, acetogenesis, and methanogenesis. In this process the complex proteins and sugars are broken down to form more simple compounds such as water, carbon dioxide, and methane.

Anaerobic digestion (Fig. 78) generates biogas with a high proportion of methane that may be used to both heat the tank and run engines or microturbines for other on-site processes. Methane generation is a key advantage of the anaerobic process. Its key disadvantage is the long time

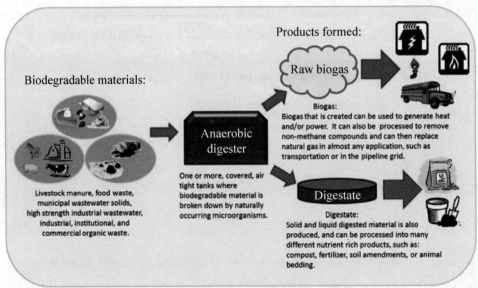

Fig. 78 Anaerobic digestion

required for the process (up to 30 days) and the high capital cost. Many larger sites utilize the biogas for combined heat and power, using the cooling water from the generators to maintain the temperature of the digestion plant at the required 35℃ ± 3℃. Sufficient energy can be generated in this way to produce more electricity than the machines require. Fig. 79 shows the process stages of anaerobic digestion.

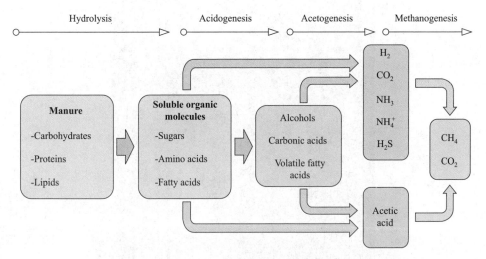

Fig. 79　Process stages of anaerobic digestion

(2) Aerobic digestion

Aerobic digestion (Fig. 80) is a bacterial process occurring in the presence of oxygen resembling a continuation of the activated sludge process. Under aerobic conditions, bacteria rapidly consume organic matter and convert it into carbon dioxide. Once there is a lack of organic matter, bacteria die and are used as food by other bacteria. This stage of the process is known as endogenous respiration. Solids reduction occurs in this phase. Because the aerobic digestion occurs much faster than anaerobic digestion, the capital costs of aerobic digestion are lower. However, the operating costs are characteristically much greater for aerobic digestion because of energy used by the blowers, pumps and motors needed to add oxygen to the process. However, recent technological advances include non-electric aerated filter sys-

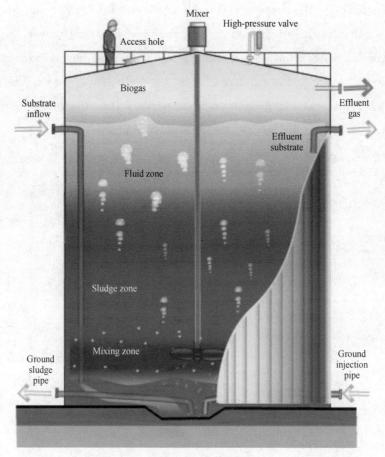

Fig. 80　Aerobic digestion

tems that use natural air currents for the aeration instead of electrically operated machinery.

　　Aerobic digestion can also be achieved by using diffuser systems or jet aerators to oxidize the sludge. Fine bubble diffusers are typically the more cost-efficient diffusion method, however, plugging is typically a problem due to sediment settling into the smaller air holes. Coarse bubble diffusers are more commonly used in activated sludge tanks or in the flocculation stages. A key component for selecting diffuser type is to ensure it will produce the required oxygen transfer rate. Fig. 81 shows the lysis of microorganisms during aerobic and anaerobic digestion.

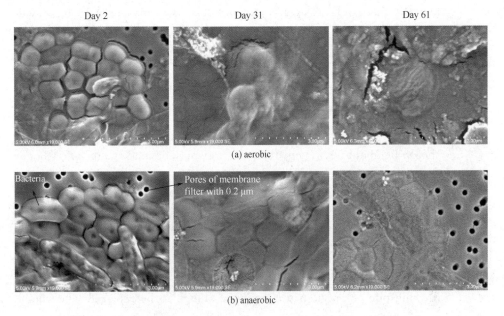

Fig. 81　Lysis of microorganisms during aerobic and anaerobic digestion

12.3.2　Outside operation practice

　　a. Comparison of anaerobic and aerobic digestion;

　　b. Aerobic and anaerobic respiration;

　　c. Digester operation.

12.4　Composting

12.4.1　Fundamentals

　　Composting is an aerobic process of mixing sewage sludge with agricultural byproduct sources of carbon such as sawdust, straw or wood chips. In the presence of oxygen, bacteria digesting both the sewage sludge and the plant materials generate heat to kill disease-causing microorganisms and parasites. Maintenance of aerobic conditions with 10%-15% oxygen requires bulking agents allowing air to circulate through the fine sludge solids. Stiff materials like corn cobs, nut shells, shredded tree-pruning waste, or bark from lumber or paper mills better separate sludge for ventilation than softer leaves and lawn clippings. Light, biologically inert bulking agents like shredded tires may be used to provide structure where small, soft plant materials are the major source of carbon.

Uniform distribution of pathogen-killing temperatures may be aided by placing an insulating blanket of previously composted sludge over aerated composting piles. Initial moisture content of the composting mixture should be about 50%; but temperatures may be inadequate for pathogen reduction where wet sludge or precipitation raises compost moisture content above 60%. Composting mixtures may be piled on concrete pads with built-in air ducts to be covered by a layer of unmixed bulking agents. Odors may be minimized by using an aerating blower drawing vacuum through the composting pile via the underlying ducts and exhausting through a filtering pile of previously composted sludge to be replaced when moisture content reaches 70%. Liquid accumulating in the underdrain ducting may be returned to the sewage treatment plant; and composting pads may be roofed to provide better moisture content control.

After a composting interval sufficient for pathogen reduction, composted piles may be screened to recover undigested bulking agents for reuse; and composted solids passing through the screen may be used as a soil amendment material with similar benefits to peat. The optimum initial carbon-to-nitrogen ratio of a composting mixture is between (26-30) : 1; but the composting ratio of agricultural byproducts may be determined by the amount required to dilute concentrations of toxic chemicals in the sludge to acceptable levels for the intended compost use. Although toxicity is low in most agricultural byproducts, suburban grass clippings may have residual herbicide levels detrimental to some agricultural uses; and freshly composted wood byproducts may contain phytotoxins inhibiting germination of seedlings until detoxified by soil fungi.

12.4.2 Key points for operation

a. Composting (Fig. 82) is as simple as:

How it works:

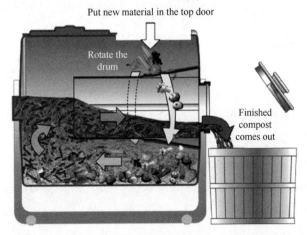

Fig. 82　Composting

　-Place materials to be composted into the sliding door on the top of the unit;

　-Turn the drum twice a week;

　-In about 6 weeks, you can start removing finished compost from the output door on the side of the unit.

　b. Product features:

　-No more waiting for batches to finish composting, continuous composter allows you to add new material and remove finished material at the same time;

　-Compost discharges automatically when the output door is opened and you turn the drum (if drum is less than half full, you'll have to open the input door on the top of the unit to remove finished compost);

　-Aerobic composting means there is little to no odor (we do not recommend adding any meat, dairy, or egg products, because these will create unpleasant odors);

　-Quick and easy assembly takes just a few minutes;

　-Easy to load and turn;

　-For best results, turn drum 3 rotations at least twice a week;

-Optional stainless steel wheels come with the unit, and can be added for easy mobility;

　　-Pest resistant.

　　c. Technical specs (One case):

　　-50 gallon capacity is ideal for composting kitchen scraps and a few cut up garden trimmings;

　　-Constructed of UV stable polyethylene (approximate 3.75 mm wall thickness);

　　-All metal parts are rust-free stainless steel;

　　-Assembled unit measures "33.5×24×31" ($L \times W \times H$)

　　-Product weight: 38 lbs;

　　-Input door height is 29 " and output door height is 16";

　　-Ships via UPS ground;

　　-One year limited manufacturer's warranty-parts replacement only.

12.5　Incineration

12.5.1　Fundamentals

　　Incineration of sludge is less common because of air emissions concerns and the supplemental fuel (typically natural gas or fuel oil) required to burn the low calorific value sludge and vaporize residual water. On a dry solids basis, the fuel value of sludge varies from about 9500 British thermal units per pound (4116J/g) of undigested sewage sludge to 2500 British thermal units per pound (1092J/g) of digested primary sludge. Stepped multiple hearth incinerators with high residence time and fluidized bed incinerators are the most common systems used to combust wastewater sludge. Co-firing in municipal waste-to-energy plants is occasionally done, this option being less expensive assuming the facilities already exist for solid waste and there is no need for auxiliary fuel. Incineration tends to maximize heavy metal concentrations in the remaining solid ash requiring disposal; but the option of returning wet scrubber effluent to the sewage treatment process may reduce air emissions by increasing concentrations of dissolved salts in sewage treatment plant effluent. Fig. 83 shows the sludge incineration process schematic.

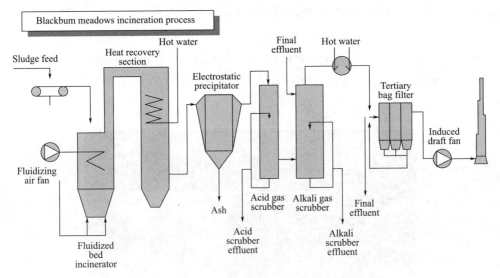

Fig. 83 Sludge incineration process schematic
(note the emphasis on air quality control)

12.5.2 Outside operation practice

a. Incineration and incinerators-in-disguise.

b. Incinerator-related issues.

-Dioxin homepage;

-Metals as catalysts for dioxin formation;

-Electrostatic precipitators breed dioxins;

-Continuous emissions monitors.

c. Incinerators-in-disguise (gasification/plasma/pyrolysis).

12.6 Microbial fuel cell

12.6.1 Fundamentals

Microbial fuel cell (MFC) is a technology that can convert organic matter into electricity through anaerobic biodegradation. Compared to other sustainable disposal methods (including composting, anaerobic digestion and carbonization), MFC has the advantage of directly recycling the cleanest energy (electricity) from organic waste with a lower generation level of secondary pollutions to water, soil and atmospheric environment. So far, many studies have been conducted using MFC to treat wastewater and organic waste (including food waste, activated sludge and animal waste). The mechanism of microbial fuel cell is shown in Fig. 84.

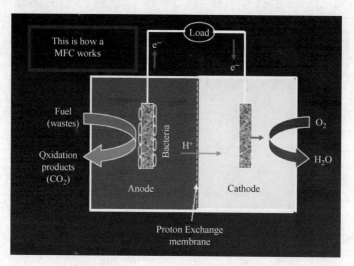

Fig. 84   Mechanism of microbial fuel cell

Fig. 85 shows the schematic diagram of simultaneous treatment of kitchen water and waste activated sludge by microbial fuel cell.

12.6.2 Key points for operation

a. Application of microbial fuel cell (Fig. 86).

b. Pollutants removal by microbial fuel cell and electricity generation (Fig. 87, Fig. 88)

12.6.3 Outside operation practice

a. Microbial fuel cell configuration and operation;

b. Assessment of electricity generation efficiency;

c. Assessment of pollutants removal efficiency.

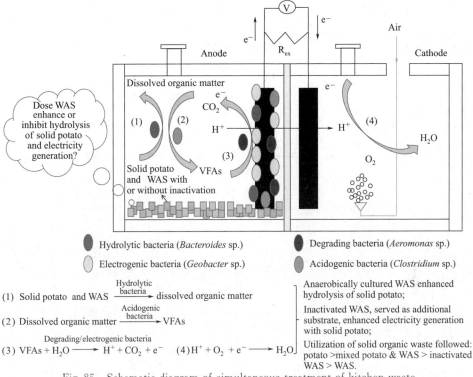

Fig. 85 Schematic diagram of simultaneous treatment of kitchen waste and waste activated sludge by microbial fuel cell

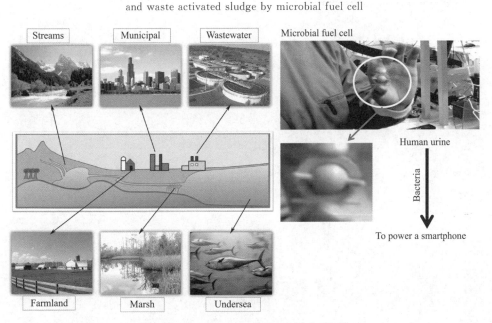

Fig. 86 Application of microbial fuel cell

Chapter 12 Sludge treatment and disposal

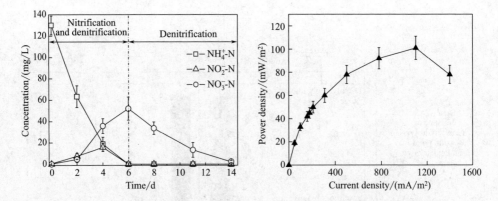

Fig. 87　Nitrogen removal and electricity generation by microbial fuel cell

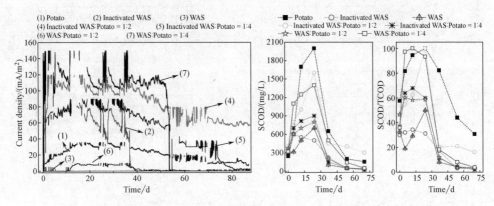

Fig. 88　Nitrogen removal and electricity generation by microbial fuel cell

Chapter 13 Advanced wastewater treatment

Primary and secondary treatment remove the majority of BOD and suspended solids found in wastewater. However, in an increasing number of cases this level of treatment has proved to be insufficient to protect the receiving water or to provide reusable water for industrial and/or domestic recycle. Thus, additional treatment steps have been added to wastewater treatment plants to provide for further organic and solids removals or to provide for removal of nutrients and/or toxic materials.

Advanced wastewater treatment will be defined as: any process designed to produce an effluent of higher quality than normally achieved by secondary treatment processes or containing unit operations not normally found in secondary treatment. The above definition is intentionally very broad and encompasses almost all unit operations not commonly found in wastewater treatment today.

(1) Types of advanced wastewater treatment

Advanced wastewater treatment may be divided into three major categories by the type of process flow scheme utilized:

    a. Tertiary treatment;

    b. Physical-chemical treatment;

    c. Combined biological-physical treatment.

(2) Advanced wastewater treatment is used for

    a. Additional organic and suspended solids removal;

    b. Removal of nitrogenous oxygen demand;

    c. Nutrient removal;

d. Removal of toxic materials;

e. Advanced wastewater treatment plant effluents may be recycled directly or indirectly to increase the available domestic water supply;

f. Advanced wastewater treatment effluents may be used for industrial process or cooling water supplies;

g. Some receiving water are not capable of withstanding the pollutional loads from the discharge of secondary effluents;

h. Secondary treatment does not remove as much of the organic pollution in wastewater as may be assumed.

13.1 Nitrogen removal

13.1.1 Fundamentals

When a treatment plant discharges into receiving stream with low available nitrogen concentration and with a flow much larger than the effluent, the presence of nitrate in the effluent generally does not adversely affect stream quality.

If the nitrate concentration in the stream is significant, it may be desirable to control the nitrogen content of the effluent, as highly nitrified effluents can still accelerate algal blooms. Even more critical is the case where treatment plant effluent is discharged directly into relatively still bodies of water such as lakes or reservoirs.

The four basic processes that are used to remove nitrogen from wastewater are:

a. Ammonia stripping;

b. Selective ion exchange;

c. Breakpoint chlorination;

d. Biological nitrification/denitrification.

13.1.2 Key points for operation

(1) Biological denitrification

Biological denitrification is a two-step process:

a. The first step is nitrification, which is conversion of ammonia to nitrate through the action of nitrifying bacteria.

b. The second step is nitrate conversion (denitrification), which is carried out by facultative heterotrophic bacteria under anoxic conditions.

Denitrification in suspended growth systems can be achieved using anyone of the typical flowsheets shown in the Fig. 89. The use of methanol or any other artificial carbon source should be avoided as far as possible for both cost and operation considerations. A more satisfactory arrangement would be to use the carbon contained in the waste itself. However, the anoxic tank has to be of sufficient detention time for denitrification to occur which, has a slower rate. If desired, a portion of the raw waste may be bypassed to enter directly into the anoxic tank and thus contribute to an increased respiration rate.

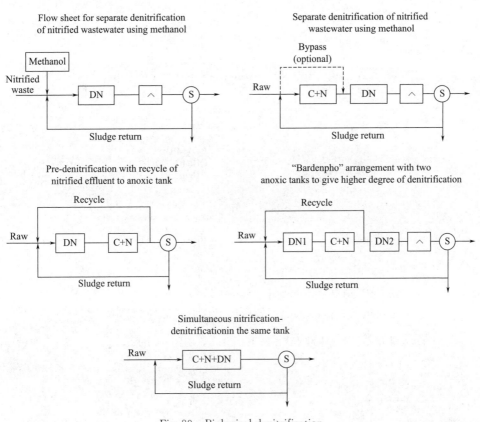

Fig. 89 Biological denitrification

By reversing the relative positions of anoxic and aerobic tanks, the oxygen requirement of the waste in its anoxic state is met by the release of oxygen from nitrates in the recycled flow taken from the end of nitrification tank. Primary settling of the raw waste may be omitted so as to bring more carbon into the anoxic tank. More complete nitrification-denitrification can be achieved by Bardenpho arrangement. The first anoxic tank has the advantage of higher denitrification rate while the nitrates remaining in the liquor passing out of the tank can be denitrified further in a second anoxic tank through endogenous respiration. The flow from anoxic tank is desirable to reaerate for 10-15 minutes to drive off nitrogen gas bubbles and add oxygen prior to sedimentation.

(2) Ammonia removal

Nitrification: ammonia (NH_3) is converted to nitrate (NO_3^-) (Fig. 90).

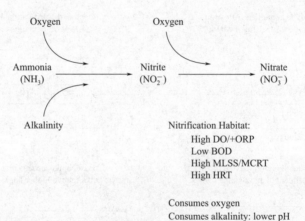

Fig. 90 Nitrification

Oxygen rich habitat:

MLSS❶ of 2500+mg/L (high sludge age/MCRT/low F: M);

ORP❶ of +100 to +150 mV (high DO);

Time❶ (high HRT...24 h, 12 h, 6 h, 4 h);

Low BOD;

---

❶ Approximate, each facility is different.

Consumes oxygen;

Adds acid-consumes 7 mg/L alkalinity per mg/L of $NH_3 \rightarrow NO_3^-$.

(3) Nitrate removal (Fig. 91)

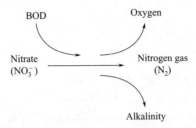

Fig. 91 Nitrate removal

a. Denitrification habitat:

Low DO/-ORP;

High BOD;

Adds DO;

Gives back ½ the alkalinity: beneficially raises pH.

b. Denitrification: Nitrate (NO_3^-) is converted to nitrogen gas (N_2).

Oxygen poor habitat:

ORP❶ of -100 mV or less (DO<0.3 mg/L);

Surplus BOD❶ (100-250 mg/L, 5-10 times as much as NO_3^-);

Retention time❶ of 45-90 minutes;

Gives back oxygen;

Gives back alkalinity (3.5 mg/L per mg/L of $NO_3^- \rightarrow N_2$).

(4) Nitrogen terms for operators

Organic-nitrogen;

Ammonia (NH_3);

Ammonium (NH_4 or NH_4^+);

TKN (Total Kjeldahl Nitrogen) = Organic-nitrogen+Ammonia;

Nitrate (NO_3 or NO_3^-);

Nitrite (NO_2 or NO_2^-);

Total nitrogen=TKN+Nitrate+Nitrite.

---

❶ Approximate, each facility is different.

13.1.3 Outside operation practice

a. Post-anoxic denitrification (Fig. 92).

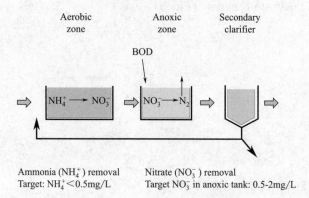

Fig. 92 Post-anoxic denitrification

b. Modified ludzack-ettinger process (Fig. 93).

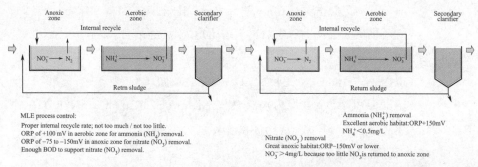

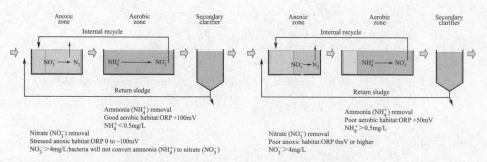

Fig. 93 Modified ludzack-ettinger process

c. Sequencing batch reactor (SBR) (Fig. 94).

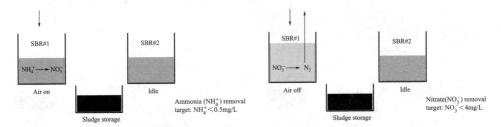

(a) SBR-Ammonia (NH_4^+) removal: nitrification

(b) SBR-Nitrate (NO_3^-) removal: denitrification

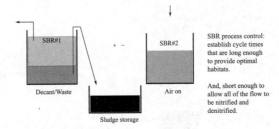

(c) Sequencing batch reactor (SBR) - Settle, decant & waste sludge

Fig. 94 Sequencing batch reactor (SBR)

d. Conventional activated sludge operated as SBR (Fig. 95).

e. Monitor and control the process.

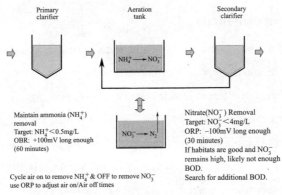

(a) MLE process modification of converntional as plant(1)

Fig. 95

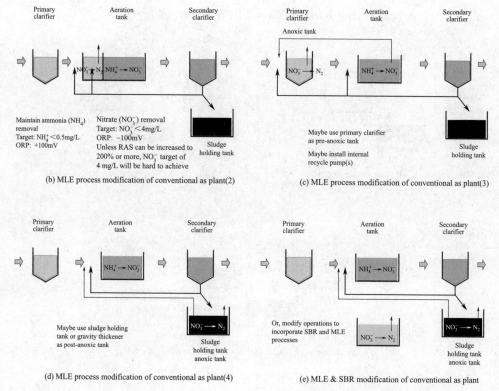

Fig. 95　Conventional activated sludge operated as SBR

13.2　Phosphorus removal

13.2.1　Fundamentals

Like nitrogen, phosphorus is also a necessary nutrient for the growth of algae. Phosphorus reduction is often needed to prevent excessive algal growth before discharging effluent into lakes, reservoirs and estuaries.

Phosphorus removal can be achieved through chemical addition and a coagulation-sedimentation process. Some biological treatment processes can also achieve phosphorus reduction, involving modifications of activated sludge treatment systems: bacteria in these systems also convert nitrate nitrogen to inert nitrogen gas and trap phosphorus in the solids that are removed from the effluent.

(1) Phosphorous removal processes

The removal of phosphorous from wastewater involves the incorporation of phosphate into TSS and the subsequent removal from these solids. Phosphorous can be incorporated into either biological solids (e. g. micro organisms) or chemical precipitates.

(2) Phosphate precipitation

Chemical precipitation is used to remove the inorganic forms of phosphate by the addition of a coagulant and a mixing of wastewater and coagulant. The multivalent metal ions most commonly used are calcium, aluminium and iron.

13.2.2 Key points for operation

(1) The main phosphate removal processes (Fig. 96)

a. Treatment of raw/primary wastewater;

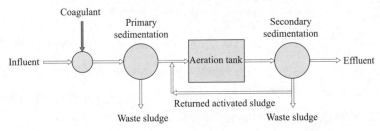

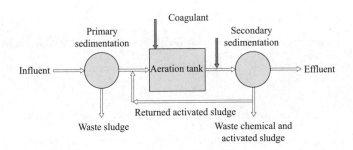

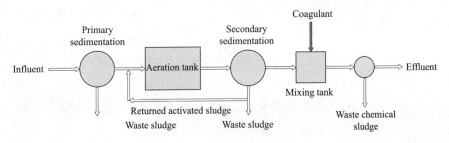

Fig. 96 The main phosphate removal processes

b. Treatment of final effluent of biological plants (postprecipitation);

c. Treatment contemporary to the secondary biologic reaction (co-precipitation).

(2) Biological phosphorus removal (Fig. 97)

The most important process is anaerobic process followed by aerobic process.

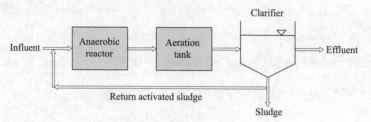

Fig. 97　Biological phosphorus removal

The principal advantages the process are reduced chemical costs and less sludge production as compared to chemical precipitation.

a. In the anaerobic zone: phosphorous accumulating organisms (PAO) assimilate fermentation products (i. e. volatile fatty acids) into storage products within the cells with the concomitant release of phosphorous from stored polyphosphates.

Acetate is produced by fermentation of COD, which is dissolved degradable organic material that can be easily assimilated by the biomass.

Using energy available from stored polyphosphates, the PAO assimilate acetate and produce intracellular polyhydroxybutyrate (PHB) storage products.

Concurrent with the acetate uptake is the release of orthophosphates, as well as magnesium, potassium, calcium cations.

b. In the aerobic zone: energy is produced by the oxidation of storage products and polyphosphate storage within the cell increases.

Stored PHB is metabolized, providing energy from oxidation and carbon for new cell growth. Some glycogen is produced from PHB metabolism.

The energy released from PHB oxidation is used to form polyphosphate bonds in cell storage. The soluble orthophosphate is removed from solution and incorporated into polyphosphates within the bacterial cell.

PHB utilization also enhances cell growth and this new biomass with high polyphosphate storage accounts for phosphorous removal.

As a portion of the biomass is wasted, the stored phosphorous is removed from the reactor for ultimate disposal with the waste sludge.

13.2.3 Outside operation practice

a. Phosphorus (P) control/removal in wastewater;

b. Impact of excess phosphorus;

c. Enhanced biological phosphorus removal.

References

[1] Neil Armitage. The challenges of sustainable urban drainage in developing countries. Urban Water Management Group University of Cape Town South Africa, 132-144.

[2] Mekonnen M M, Hoekstra A Y. National water footprint accounts: the green, blue and grey water footprint of production and consumption, Value of Water Research Report Series No. 50, UNESCO-IHE, Delft, the Netherlands. Water Footprint Network. 2011. Archived from the original on 2014.

[3] World population to reach 9.1 billion in 2050, UN projects, Un. org. 2005-02-24. Retrieved 2009-03-12.

[4] Burden of disease and cost-effectiveness estimates, World Health Organization. Retrieved April 5, 2014.

[5] Environmental quality standards for surface water, GB 3838—2002.

[6] Envirofacts United States Environmental Protection Agency (US EPA).

[7] Jeff McMahon. EPA Draft Stirs Fears of Radically Relaxed Radiation Guidelines, Forbes. Retrieved 2013-05-14.

[8] Markham V D, Steinzor N. U. S. national report on population and the environment. New Canaan Connecticut Center for Environment and Population, 2006, 43 (2), 433-442.

[9] Pamphlet (or booklet). Quarantine and isolation: lessons learned from sars: a report to the centers for disease control and prevention, 2003.

[10] A Koch Chemical Technology Group. LLC Company. Memerane Systems.

[11] Fusheng Li, Xuemei Tan, Yongfen Wei. Disinfection efficacy of waterborne enteric viruses under low chlorine doses: Investigation based on infectivity and genomic integrity of a viral surrogate, in Water Supply and Water Quality, eds. Jan F. Lemanski and Sergiusz Zabawa, Polskie Zrzeszenie Inzynierow I Technikow Sanitarnych Oddzial Wielkopolski (Publisher), 2014, 801-812.

[12] Fusheng Li, Hongjie Gui, Haixia Du, et al. Optimum combination of existing treatment processes is essential for elevation of drinking water quality, in Water Supply and Water Quality, eds. Zbyslaw Dymaczewski and Joanna Jez-Walkowiak & Andrzej Urbaniak, polskie Zrzeszenie Inzynierow i Technikow Sanitarnych Oddzial Wielkopolski (Publisher), 2016, 623-641.

[13] Haixia Du, Fusheng Li. Enhancement of solid potato waste treatment by microbial fuel cell with mixed feeding of waste activated sludge. Journal of Cleaner Production, 2017, 143: 336-344.

[14] Haixia Du, Fusheng Li. Effect of increasing the mass fraction of boiled potato in the potato feed on the performance of microbial fuel cell. Science of the Total Environment, 2016, 569: 841-849.

[15] Haixia Du, Fusheng Li. Characteristics of dissolved organic matter formed in aerobic and anaerobic digestion of excess activated sludge. Chemosphere, 2017, 168: 1022-1031.

[16] Haixia Du, Fusheng Li. Size effects of potato waste on its treatment by microbial fuel cell. Environmental Technology, 2016, 37 (10): 1305-1313.
[17] Haixia Du, Fusheng Li, Zaiji Yu, Chunhua Feng, Wenhan Li. Nitrification and denitrification in two-chamber microbial fuel cells for treatment of wastewater containing high concentrations of ammonia nitrogen. Environmental Technology, 2016, 37 (10): 1232-1239.
[18] Jachens R C, Holzer T L. Differential compaction mechanism for earth fissures near casa grande, arizona. Geological Society of America Bulletin, 1982, 93 (10): 998-1012.
[19] Chitjian M, Koizumi J. Parsons/Engineering Science, 2000.
[20] United States Environmental Protection Agency (US EPA). 2000. Oxidation Ditch. Wastewater Technology Fact Sheet.
[21] Wastewater Technology Fact Sheet Trickling Filters, EPA 832-F-00-014, September 2000. United States Environmental Protection Agency.
[22] Haixia Du, Fusheng Li. Characteristics of dissolved organic matter formed in aerobic and anaerobic digestion of excess activated sludge. Chemosphere, 2017, 168: 1022-1031.
[23] Metcalf, Eddy. Wastewater Engineering, Treatment, Disposal and Reuse. McGraw Hill, New York, 1997.
[24] Turovskiy I S, Mathai P K. Wastewater Sludge Processing. A John Wiley & Sons, Inc., Publication, 2006.